A HISTORY OF
THE CONCEPT OF VALENCY

A HISTORY OF
THE CONCEPT OF VALENCY
TO 1930

W. G. PALMER
M.A., Sc.D., D.Sc.

Fellow of St John's College, Cambridge,
formerly University Lecturer in Chemistry in
the University of Cambridge

CAMBRIDGE
AT THE UNIVERSITY PRESS
1965

PUBLISHED BY
THE SYNDICS OF THE CAMBRIDGE UNIVERSITY PRESS

Bentley House, 200 Euston Road, London, N.W.1
American Branch: 32 East 57th Street, New York, N.Y. 10022
West African Office: P.O. Box 33, Ibadan, Nigeria

©

CAMBRIDGE UNIVERSITY PRESS

1965

Printed in Great Britain at the University Printing House, Cambridge
(Brooke Crutchley, University Printer)

LIBRARY OF CONGRESS CATALOGUE
CARD NUMBER: 65–14348

CONTENTS

PREFACE

The following pages constitute an expansion of a short course of lectures delivered a few years ago in the University of Cambridge on the invitation of the Committee for the History and Philosophy of Science.

Throughout the book I have constantly endeavoured to preserve the historical outlook of different chronological stages by presenting extracts from the publications of the most influential authorities, largely in their own languages and in the nomenclature and terminology of their time. In this I am happy to follow the lead of a number of contemporary publications on the history of science. I hope that the translations I have occasionally appended to such excerpts may earn the approval of readers, and especially of those who, not being professional scientists, might otherwise find themselves baffled by a foreign terminology.

Since the history of science would be devoid of interest or indeed meaningless if it did not include in its scope something of the personalities and circumstances at least of those who conceived the principal advances I need offer no excuses for including biographical notes and references.

By 1930 the discovery of wave-mechanics had provided a firm theoretical basis for the arbitrary postulates upon which conceptions of atomic structure had previously rested, and it became possible for the first time to give a better than empirical explanation of chemical union, and consequently of the principle of valency. It is however yet too early to assess the very rapid developments since 1930 in a just historical perspective.

May, 1964   W. G. P.

LIST OF
ABBREVIATED TITLES

Liebigs Ann.	*Liebig's Annalen der Chemie*
Ann. Chim. (Pharm.)	*Annales de Chimie et de Pharmacie*
Ann. Chim. (Phys.)	*Annales de Chimie et de Physique*
Ann. Phys., Lpz.	*Annalen der Physik, Leipzig*
Ann. Phys., Paris	*Annales de Physique, Paris*
Liebigs Ann. (Suppl.)	Supplements to *Liebig's Annalen der Chemie*
Ber. dtsch. chem. Ges.	*Berichte der Deutschen Chemischen Gesellschaft*
Bull. Acad. Belg. Cl. Sci.	*Bulletin de l'Académie royale de Belgique*
Bull. Soc. chim. Fr.	*Bulletin de la Société chimique de France*
C.R. Acad. Sci., Paris	*Comptes rendus hebdomodaires des Séances de l'Académie des Sciences*
Gazz. chim. ital.	*Gazzetta chimica italiana*
Helv. chim. acta	*Helvetica Chimica Acta*
J. Amer. chem. Soc.	*Journal of the American Chemical Society*
J. chem. Soc.	*Journal of the Chemical Society*
J. prakt. Chem.	*Journal für praktische Chemie*
J. Soc. phys.-chim. russ.	*Journal of the Russian Physical and Chemical Society*
Phil. Mag.	*Philosophical Magazine*
Phil. Trans.	*Philosophical Transactions of the Royal Society*
Phys. Rev.	*Physical Review*
Phys. Z.	*Physikalische Zeitschrift*
Poggendorfs Ann.	*Poggendorf's Annalen*
Proc. chem. Soc., Lond.	*Proceedings of the Chemical Society, London*
Proc. roy. Soc.	*Proceedings of the Royal Society*
Proc. roy. Soc. Edinb.	*Proceedings of the Royal Society of Edinburgh*
Quart. J. chem. Soc.	*Quarterly Journal of the Chemical Society* (1849–1862 incl.)
Trans. Faraday Soc.	*Transactions of the Faraday Society*
Trans. roy. Soc. Edinb.	*Transactions of the Royal Society of Edinburgh*
Verh. dtsch. phys. Ges.	*Verhandlungen der Deutschen Physikalischen Gesellschaft*
Z. Chem.	*Zeitschrift für Chemie* (Göttingen, 1865–1871, vols. 1–7)
Z. Elektrochem.	*Zeitschrift für Elektrochemie*
Z. Kristallogr.	*Zeitschrift für Kristallographie*
Z. Phys.	*Zeitschrift für Physik*
Z. phys. Chem.	*Zeitschrift für physikalische Chemie*

INTRODUCTION

We have a universally acknowledged system of constitutional formulae based on valency, which we define as the habit in regard to combination exhibited by elementary atoms, without necessarily forming any hypothesis on the nature of affinity.

SIR WILLIAM TILDEN, Cannizzaro Memorial Lecture
of the Chemical Society, 1912

Today more than one hundred chemical elements are recognized, and it is known that almost all of them form compounds, or fluorides, of the general chemical formula AF_n, with the element fluorine (symbol F). The known formulae of these compounds provide, however, no examples of the type A_mF in which the number m is greater than unity. The number n derived from the formulae AF_n therefore affords a genuine measure of the extent to which the total combining power of the element A is subdivided, since the combining power of fluorine is evidently indivisible: and it defines the *valency* of the element A towards fluorine.

Although it might well be expected on general grounds that each element would maintain a fixed but specific valency in all its compounds, only some thirty elements have this property. The valencies of those elements, carbon, nitrogen, oxygen, and hydrogen, with the multifarious combinations of which organic chemistry is primarily concerned, are invariable (C, IV; N, III; O, II; H, I) and consequently organic chemists, by devising consistent structural formulae based on these fixed valencies, achieved spectacular advances. It is among the metallic elements that the most numerous examples of varying valency are to be found: it was fortunate that Frankland (in 1852) based the earliest explicit formulation of the concept of valency upon his studies of (organo-) metallic compounds, and indicated from the first the possibility of variable valency. Many chemists, and especially Kekulé, who welcomed and exploited the concept with great acumen and enthusiasm, found it repugnant to follow Frankland's lead, and thereby entailed continued confusion and uncertainty in the field of inorganic chemistry. Indeed it was only in the present century when the then well-established principle of atomic valency became the main incentive towards a general theory of chemical combination (in which the principle became merged) that such difficulties were finally resolved.

The concluding clause of the excerpt from Tilden's lecture must almost certainly be the last public example of the over-complacency or even pride of the chemists of the later nineteenth century in the empiricism of their science, in which they differed widely from the contemporary physicists. The chemical attitude was no doubt engendered and fostered by the supreme successes of organic chemistry, resting only on the empirical principles of valency. Less than a year after the delivery of Tilden's lecture Bohr was to give his first exposition of the 'Rutherford-Bohr' model of the atom, to be followed in 1916 by the tentative but extremely significant speculations of G. N. Lewis on the nature of chemical union and the explanation of valency. The modern historian of chemistry must therefore regard the doctrine of valency as destined from the first to be subsumed in a general principle of chemical union. It is for this reason not irrelevant to review briefly the ideas on chemical combination which had gained wide acceptance before the rise of a clear conception of valency, and which were undoubtedly in the minds of those, including Frankland, who promulgated and first exploited its usefulness.

Scarcely any chemist at the present day speaks of Radicals or of Types: and yet the type-theory, like its forerunner, the dualistic view, had its good points. It would, I think, be wise, not wholly to lose sight of these several theories, which, after all, are based upon a considerable number of facts.*

* Kekulé, 'On some points of chemical philosophy', *The Laboratory*, 1 (1867), 303.

1

DALTON AND WOLLASTON

Early notions of chemical combination

Chemical analysis or synthesis can go no further than to the separation of particles one from another, and to their reunion. No new creation or destruction of matter is within the reach of chemical agency. We might as well attempt to introduce a new planet into the solar system, or to annihilate one already in existence, as to create or destroy a particle of hydrogen. All the changes we can produce consist in separating particles that are in a state of cohesion or combination, and joining those that were previously at a distance.

JOHN DALTON, *A New System of Chemical Philosophy*, vol. 1, part 1 (1808), p. 212.*

John Dalton (1766-1844)

John Dalton was born in 1766 at Eaglesfield, near Cockermouth in Cumberland, into a family of small proprietors. His paternal grandfather had joined the Society of Friends and his grandson was undoubtedly much influenced in his habits and character by the continuance of his parents in that body. He attended the village school, but only during the winter months, and in later life testified to the salutary effect of the outdoor exertions, willingly entered upon with his school mates during the months of summer, in fortifying the robust health he enjoyed all his life. Such was his progress, especially in 'ciphering', that at the age of twelve he took part in the teaching, and a year or two later enabled his former master to retire.

A cousin, George Bewley, kept a school in Kendal, where Dalton's elder brother Jonathan had already become an assistant, when John at the age of fifteen joined him. In 1785

---

* Cited hereafter as *New System*.

Bewley resigned the school to the brothers Dalton. It was during his twelve years at Kendal that Dalton found in John Gough (also a member of the Society of Friends) a scientific preceptor of whom he afterwards wrote 'For about eight years we were intimately acquainted. Mr. Gough was as much gratified with imparting his stores of science as I was in receiving them.... It was he who first set the example of keeping a meteorological journal.'*

Gough, who lost his sight in infancy, also prepared Whewell for his entry at Cambridge in 1812, and stirred him to write 'a blind man, but very eminent in classics, mathematics, botany, and chemistry': and Wordsworth acknowledged that the remarkable portrait in Book VII (ll. 482–515) of *The Excursion* was drawn from Gough. It was undoubtedly through Gough that Dalton was introduced to the writings of Newton, from which he formed his conviction of the atomic nature of matter. Dalton is said to have taken during his life upwards of 40,000 meteorological observations, the last on the day he died. He himself constructed the barometers and thermometers necessary for the purpose, and it may be remembered that it was an essentially meteorological problem that led to his first published table of atomic weights (p. 87).

With an intention, it may be thought, to clarify and consolidate his own knowledge in imparting it to others, in October 1787 Dalton circulated in Kendal a notice: 'Twelve Lectures on Natural Philosophy to be read at the school (if a sufficient number of subscribers are procured) by John Dalton.' The syllabus contained sections on 'Mechanics, Optics, Pneumatics, Astronomy, and the Use of the Globes' and concluded with 'Ex rerum causis supremam noscere causam'. It was again Gough who in 1793 launched Dalton into wider fame, by recommending him as a teacher of mathematics in the New College of Manchester, which had arisen out of the Warrington Academy, where Priestley had taught. The teaching at Manchester was evidently more catholic than its description suggests for Dalton used Lavoisier's new *Elements of Chemistry* in his work. Although

* Preface, *Meteorological Observations and Essays*, 2nd edition (London, 1834).

making his first visit to so large a town Dalton was soon recognized by being elected to membership of the Literary and Philosophical Society in 1794, and a month after his election read his first communication to the society (on his colour-blindness). He became Secretary in 1800 and was President of the Society from 1817 until his death in 1844. One of his early Memoirs which recorded his experiments leading to the observation that 'elastic fluids are equally expanded by heat', was published six months before Gay-Lussac's similar observations. Dalton had in early life developed a habit of extreme diligence in all forms of intellectual activity, and replied to an inquiry, that he had never found the time to marry.

We must see him (Dalton) as a man having limited assistance from the knowledge of others, ignorant of many of the elegant and easy methods of procuring knowledge and illustrating facts which have become the common inheritance of universities, and accustomed to the society of few only who had similar studies....Davy and Dumas add poetry and eloquence to sterling scientific vigour, whilst in Dalton we find only strength and rude simplicity.*

During the early nineteenth century the term 'gas' was used more specifically than today: for example, nitrous gas (nitric oxide), olefiant gas (ethylene). When it was desired to focus attention on the gaseous state in general the term 'elastic fluid' was commonly preferred. In this state the minute units of matter were thought to be embedded in a voluminous and 'globular' envelope of 'caloric' or heat. The elastic properties were ascribed to the compressibility of this envelope, and Boyle's law was explained by supposing that through its agency the chemical units repelled each other inversely as the distance between them.

When any body exists in the elastic state, its ultimate particles are separated from each other to a much greater distance than in any other state; each particle occupies the centre of a comparatively large sphere, and supports its dignity by keeping all the rest, which by their gravity, or otherwise, are disposed to encroach up to it, at a respectful distance.†

* From 'Memoir of Dr Dalton', by Dr Angus Smith, Secretary of the Manchester Literary and Philosophical Society, 1856.
† Dalton, *New System*, vol. I, part I, p. 211.

On withdrawal of heat the gaseous system contracted, liquefaction and finally solidification ensuing. To a world accepting a kinetic theory of matter as almost axiomatic such an entirely static concept may seem incredible, but before the general recognition of 'heat as a mode of motion' (under the lead of Clausius and Clerk Maxwell from 1857) it gave a workable, and it may be the only possible model.

Respecting the nature of the principle (heat) there is diversity of sentiment: some supposing it a *substance*, others a *quality* or property of substance. Boerhaave, followed by most moderns, is of the former opinion; Newton with some others, are of the latter: these conceive heat to consist in an internal vibratory motion of the particles of bodies.*

The problem of the spontaneous interdiffusion of gases presented very serious difficulties, and Dalton, especially in his role as a meteorologist concerned with the atmospheric gaseous mixture, gave much attention to it. His discovery of the law now universally accepted as 'Dalton's law of partial pressures', which asserted that from a mixture of 'elastic fluids' each constituent dissolves in a liquid as though the others were absent, led him to the belief that in such a mixture only the *like* 'particles' repel each other: for in the process of solution like particles appear to suffer no constraint (or pressure) from the admixed unlike particles.

Thus for Dalton the physical properties of gases were related to a natural repulsion between like particles. It appeared to follow that *attractive* force was to be expected only between unlike particles and this Dalton identified with chemical affinity, leading to 'chemical synthesis'. Herein lay a main cause of his antagonism to Avogadro's thesis, which required *like* particles to exhibit attraction and chemical union (to give 'elementary molecules'); and also the central theme to be embodied somewhat later by Berzelius in his dualistic theory. The (compound) atoms resulting from 'synthesis' were classified as follows:

If there are two bodies *A* and *B*, which are disposed to combine, the following is the order in which the combinations may take place, beginning with the most simple: namely,

* Dalton, *Meteorological Observations and Essays*, 2nd edition (1834), p. 17.

6

1 atom of A + 1 atom of B = 1 atom of C, *binary*,
1 atom of A + 2 atoms of B = 1 atom of D, *ternary*,
2 atoms of A + 1 atom of B = 1 atom of E, *ternary*,
1 atom of A + 3 atoms of B = 1 atom of F, *quaternary*,
3 atoms of A + 1 atom of B = 1 atom of G, *quaternary*, etc. etc.*

Dalton's profession as a schoolmaster was probably not unrelated to the somewhat abrupt style in which he was apt to publish his chemical principles, with little discussion and as almost self-evident. He could, however, be stirred by criticism to a more lively and persuasive vein. Such an occasion occurred in 1811 when, in a letter to Nicholson's Journal,† one of the principal scientific journals of the day and a forerunner of the 'Philosophical Magazine', Dr John Bostock, a practising chemist of some distinction, challenged Dalton's rules of chemical combination as arbitrary. These rules, published in the *New System*,‡ were as follows:

1st. When only one combination of two bodies can be obtained, it must be presumed to be a *binary* one, unless some cause appear to the contrary.

2nd. When two combinations are observed, they must be presumed to be a *binary* and a *ternary*.

3rd. When three combinations are obtained, we may expect one to be a *binary*, and the other two *ternary*.

4th. When four combinations are observed, we should expect one *binary*, two *ternary* and one *quaternary*, etc.

5th. A *binary* compound should always be specifically heavier than the mere mixture of its two ingredients.

6th. A *ternary* compound should be specifically heavier than the mixture of a binary and a simple, which would, if combined, constitute it.

7th. The above rules and observations apply, when two compound bodies are combined.

The following passages are extracted from Dalton's long and reasoned reply to Bostock, in Nicholson's Journal.§

I proceed now to point out the mechanical consistency of the 1st rule, which Dr Bostock has quoted, p. 283, namely, that 'when only one combination of two bodies can be obtained, it must be presumed to be a *binary* one, unless some cause appear to the contrary'; and if this be established, the other three he quotes may be considered as corollaries from it.

* *New System*, vol. 1, part 1, p. 213.
† Vol. **28**, p. 280. The editor and proprietor, Wm. Nicholson, F.R.S., was, with Carlisle, the first to electrolyse water.
‡ Vol. 1, part 1, p. 214.   § Vol. **29**, p. 145 (1811).

Let us suppose a mixture, for instance, of hidrogenous and oxigenous gas, in such sort, that there are the same number of atoms of each gas; now as the gasses are uniformly diffused, each atom of hidrogen must have one of oxigen more immediately in its vicinity. The atoms of hidrogen are all repulsive to each other; so are those of oxigen; the atoms of hidrogen are all equally attractive of those of oxigen, and the attraction increases in some unknown ratio as the distance diminishes. Heat, or some other power prevents the union of the two elements, till by an electric spark, or some other stimulus, the equilibrium is disturbed, when the power of affinity is enabled to overcome the obstacles to its efficiency, and a chemical union of the elementary particles of hidrogen and oxigen ensues.

In the next paragraph Dalton contrasts the case that each hydrogen atom combines with its neighbouring oxygen ('and no part of either elementary gas is left uncombined') with the supposition that an oxygen atom 'should leap over the more proximate atoms of hidrogen to another at a greater distance, and consequently less attractive'.

The 2d, 3d, and 4th rules are necessarily consequent to the 1st. When an element A has an affinity for another B, I see no mechanical reason why it should not take as many atoms of B as are presented to it and can possibly come into contact with it (which may probably be 12 in general), *except so ar as the repulsion of the atoms of B among themselves are more than a match for the attraction of an atom of A.* Now this repulsion begins with 2 atoms of B to one of A, in which case the 2 atoms of B are diametrically opposed; it increases with 3 atoms of B to one of A, in which case the atoms of B are only 120° asunder; with 4 atoms of B it is still greater as the distance is then only 90°; and so on in proportion to the number of atoms.[1]*

[1] [Footnote] I find from the principles of statics, that, upon the supposition of spherical atoms of equal size, and that the law of repulsion after chemical union is the same as before, namely, reciprocally as the central distance, the repulsion of any one atom of B upon another of B, to separate it from A, is a constant quantity, on whatever point of the surface of A it may be placed; so that when there are three atoms of B the 3d atom is repelled twice as much by the other two as it would be by a single atom placed diametrically opposite. When there are 4 atoms, then the 4th is three times as much repelled, &c.

It is evident then from these positions, that, as far as powers of attraction and repulsion are concerned, (and we know of no other in chemistry) *binary* compounds must first be formed in the ordinary course of things, then *ternary*, and so on, till the repulsion of the atoms of B (or A whichever happens to be on the surface of the other), refuse to admit any more.

I must object to such loose quotations as the following; namely that I have assumed, 'that when only one compound of two elementary bodies can be obtained, it must be binary'; my language is, 'it must be *presumed to be* a binary one *unless some cause appear to the contrary*'.†

* Vol. **29**, p. 147 (1811).   † *Ibid.* p. 149.

[Dalton then quotes from his *New System*] 'As only *one* compound of oxigen and hidrogen is certainly known,' and 'only one compound of hidrogen and azote has yet been discovered'.

The 5th rule is that a binary compound should always be specifically heavier than the mere mixture of its two ingredients. The principle on which this rule is founded is recognized by chemists as *general*, if not *universal*; namely, that condensation of volume is a necessary consequence of the expulsion of heat by the exertion of affinity. Thus steam is specifically heavier than a mixture of 2 parts hidrogen and 1 oxigen; ammoniacal gas is in like manner heavier than 21 azote with 72 hidrogen.*

It is remarkable that Dalton should nowhere else in his writings refer to a 'principle recognized as general, if not universal', nor does he appear to have expected a significant relation between the magnitude of the 'expulsion of heat' and the strength of the affinity exerted. By the time of Dalton's death in 1844 trustworthy data in thermochemistry were being actively accumulated, and about twenty years later the two principal pioneers in this field, Thomsen (in Copenhagen) and M. P. E. Berthelot (in Paris) put forward the theorem that the heat of chemical reaction directly measured the chemical affinity of the reactants, a principle afterwards shown to be an over-simplification. As will be seen in the next chapter speculations upon the possible source of the heat released in chemical action formed one of the bases of Berzelius' dualistic theory.

During the year 1808 one of Dalton's most eminent English contemporaries, Wm. Hyde Wollaston (1766–1828), had published a memoir entitled 'On super-acid and sub-acid salts'†, in which, from accurate and ingeniously contrived experiments, he demonstrated that in the 'sulphates of potash' and the 'carbonates of potash and soda' the ratios of equivalents of base to acid are exactly 1 : 1 and 1 : 2 respectively. The same relation was found between the 'neutral oxalate' and the 'binoxalate' of potash ($K_2C_2O_4$ and KHC_2O_4 in modern formulae), but when one equivalent of potash and three equivalents of oxalic acid were brought together the result was a mixture in equal proportions of two salts, the one the 'binoxalate' with proportions

* *Ibid.* p. 148.
† *Phil. Trans.* **98**, 96 (1808).

1:2 and the other a new salt, termed 'quadroxalate', in proportions 1:4.

To account for this want of disposition to unite in the proportion of three to one by Mr Dalton's theory, I apprehend he might consider the neutral salt as consisting of

2 particles potash with 1 acid

The binoxalate

    as 1 and 1, or 2   with 2,

The quadroxalate

    as 1 and 2, or 2   with 4,

in which cases the ratios which I have observed of the acids to each other in these salts would respectively obtain.

But an explanation, which admits the supposition of a double share of potash in the neutral salt, is not altogether satisfactory; and I am inclined to think, that when our views are sufficiently extended, to enable us to reason with precision concerning the proportions of elementary atoms, we shall find the arithmetical relation alone will not be sufficient to explain their mutual action, and that we shall be obliged to acquire a geometrical conception of their relative arrangement in all three dimensions of solid extension.

For instance, if we suppose the limit to the approach of particles to be the same in all directions, and hence their virtual extent to be spherical (which is the most simple hypothesis); in this case when different sorts combine singly there is but one mode of union. If they unite in the proportion of two to one, the two particles will naturally arrange themselves at opposite poles of that to which they unite. If there be three, they might be arranged with regularity, at the angles of an equilateral triangle in a great circle surrounding the single spherule; but in this arrangement, for want of similar matter at the poles of this circle, the equilibrium would be unstable and would be liable to be deranged by the slightest force of adjacent combinations; but when the number of one set of particles exceeds in the proportion of four to one, then, a stable equilibrium may take place, if the four particles are situated at the angles of the four equilateral triangles composing a regular tetrahedron.

It is perhaps too much to hope, that the geometrical arrangement of primary particles will ever be perfectly known; since even admitting that a very small number of these atoms combining together would have a tendency to arrange themselves in the manner I have imagined; yet, until it is ascertained how small a proportion the primary particles themselves bear to the interval between them, it may be supposed that surrounding combinations, although themselves analogous, might disturb that arrangement, and in that case, the effect of such interference must also be taken into account, before any theory of chemical combination can be rendered complete.

Wollaston's perception in 1808 that four identical 'particles' would tend to adopt a tetrahedral arrangement around a fifth particle, and not that of a plane square as Dalton had assumed, was to be forgotten for nearly 70 years; but his implied opinion that Dalton's rules were based upon a too simple conception of the nature of chemical union was already held by many chemists.

Assuming that the atoms of elementary bodies are spherical Dalton included in the *New System* the earliest graphical formulae, some examples of which are represented below, together with Dalton's legends:

Hydrogen   Oxygen   Azote*

Water   Ammonia   Nitrous gas

Nitrous oxide   Nitric acid

Nitrous acid
(1 nitric acid
+ 1 nitrous gas)

Nitrate of ammonia
(1 nitric acid
+ 1 ammonia + 1 water)

---

* The name *azote*, first suggested by Lavoisier, de Fourcroy and de Morveau in *Méthode de Nomenclature chimique* (Paris, 1787), persisted, with its symbol Az, in French chemical literature into the present century. J. A. Chaptal proposed 'nitrogen', in analogy with oxygen, in 1790.

His exposition of such formulae (which he retained unchanged in the last edition of the *New System* in 1842) reveals little doubt that in his mind they each portrayed a real physical entity, a character to be so strenuously and so tediously denied to their structural formulae by the later nineteenth-century chemists.

It is perhaps not generally known that Dalton also made use in his teaching of molecular models similar to those so often seen on present-day lecture tables. During 1842 he privately published his last scientific papers: 'On phosphates and arsenates', 'Acids, bases and water', 'A new and easy method of analysing sugar'.* The following matter forms part of the last paper:

It is my opinion that the simple atoms are *alike*, *globular* and all of the same *magnitude* or *bulk*, whether of hydrogen 1 or lead 90. My friend Mr Ewart, at my suggestion, made me a number of equal balls, about an inch in diameter, about 30 years ago; they have been in use ever since, I occasionally showing them to my pupils. One ball had 12 holes in it, equidistant; and 12 pins were stuck in the other balls, so as to arrange the 12 around the one and be in contact with it: they (the 12) were about 1/10 inch asunder. Another ball, with 8 equidistant holes in it: and they (the 8) were about 3/10 of an inch asunder, a regular series of equidistant atoms. The 7 are an awkward number to arrange around 1 atom. The 6 are an equidistant number of atoms, 90° asunder, 2 at the poles, and 4 at the equator. The 5 are a *symmetrical* number, 2 at the poles, and three at the equator: but the latter are 120° asunder, and the 2 at the poles are 90° from the equator. The 4 is a split of the 8, a regular number and equidistant. The 3 are around the equator at 120° asunder; the 2 at opposite poles. I had no idea at that time (30 years ago) that the atoms were all of a *bulk*; but for the sake of illustration I had them made alike.

There is an obvious incompatibility in the description regarding the 12-co-ordinated (close-packed) model, but when the essay was published in 1842 Dalton's memory of details was probably failing, following the illness of 1837 from which he never fully recovered. It is not unlikely that the maker of the models, with foresight to their easier manipulation, provided rather larger balls for the central position.

It is curious that Henry in his *Life of Dalton*† discusses *in extenso* Dalton's last two papers and the confirmation of much of their experimental material by Playfair and Joule, but completely overlooks the description of his models, which might well

* Harrison, Manchester, 1842.   † Cavendish Society, 1854, pp. 193–7.

carry more interest for a latter-day chemist than the main subject-matter.

From the position of an unrivalled innovator in chemical thought during his years of early middle age Dalton, without relinquishing his high prestige, receded behind his younger contemporaries who, proud to regard themselves as his disciples, were actively prosecuting, expanding and therefore inevitably modifying his formulation of the atomic theory. Amongst these, thirteen years Dalton's junior but frequently corresponding with him, Berzelius stood pre-eminent. In a letter dated 16 October 1812 he strongly urges Dalton to re-consider his total rejection of the experiments and ideas of Gay-Lussac and Avogadro upon the combining volumes of gases. To this Dalton replied four days later: 'The French doctrine of equal measures of gases combining etc. is what I do not accept, understanding it in a mathematical sense. At the same time I acknowledge there is something wonderful in the frequency of the approximation.' One can perhaps detect a certain air of wistfulness in the last sentence. Nevertheless, in his *New System*, vol. II, part I, which appeared in 1827, Dalton still maintains* an unmitigated scepticism:

The combinations of gases in equal or multiple volumes, or at least approximations to them frequently occur: but till a principle has been suggested to account for the phenomenon I think we ought to investigate the facts with great care and not suffer ourselves to be led to adopt analyses till some reason can be discovered for them.

The apparent denial to chemistry of the benefits of scientific induction coupled with the veiled impugnment of Gay-Lussac's experiments could hardly fail to place Dalton in an isolated position. In the 1807 edition of his *System of Chemistry*, Dalton's devoted friend Thomas Thomson, Professor at Glasgow, had been allowed and even encouraged to publish the first account of the new atomic theory. In a subsequent edition of 1825 Thomson wrote:

When he (Dalton) did see it [Gay-Lussac's account of his law of gaseous combination, in 1808] he set himself to show that Gay-Lussac's opinions

* Appendix, p. 349.

13

were ill-founded. But the later researches of chemists have left no doubts about their accuracy; and if Mr Dalton still withholds his assent, he is, I believe, the only living chemist to do so.

The unrelenting rejection of the 'French doctrine' and consequently of Avogadro's explanation of it, by the re-founder of the atomic theory was probably the most powerful single influence maintaining that state of baleful confusion in the field of atomic and molecular weights which was ended only in 1860 by the successful effort of Cannizzaro to re-instate the views of his countryman at the Karlsruhe Congress held in that year. Of the effect upon the discovery and application of the doctrine of valency Frankland wrote:*

In claiming the discovery of the law of atomicity in the study of organo-metallic bodies...I do not forget how much this law in its present development owes to the labours of Cannizzaro. Indeed until the latter had placed the atomic weights of metallic elements upon their present consistent basis, the satisfactory development of the doctrine was impossible.

In referring to Dalton's intransigent attitude to Gay-Lussac's generalization Dr Angus Smith, who at the request of the Manchester Literary and Philosophical Society of which he was Secretary, prepared a Memoir of its President, which appeared in 1856, felt compelled to write 'He is strict in the examination of the analyses of others, and seeks mathematical precision, when he could so very easily, in his own researches, overleap a few per cent.'. Berzelius earlier made the comment:

Dans son nouveau système de chimie, Dalton vient d'examiner les corps oxidés, et il indique le nombre d'atomes qu'il suppose y être contenus. Il paraît cependant que, dans ce travail, ce savant distingué s'est trop peu fondé sur l'expérience, et peut-être n'a-t-il pas agi avec assez de précaution en applicant la nouvelle hypothèse au système de la chimie. Il m'a semblé que, dans le petit nombre d'analyses qu'il a publiées, l'on pouvait quelquefois s'apercevoir du désir de l'opérateur d' obtenir un certain résultat; ce dont on ne peut trop se garder lorsqu'on cherche des preuves pour ou contre une théorie dont on est préoccupé.†

* *Collected Works*, p. 154.
† *La théorie des proportions chimique et de l'influence de l'électricité dans la nature inorganique*; authorized French edition (Paris, 1835). Cited hereafter as *Théorie*.

BIOGRAPHICAL AND OTHER REFERENCES

A New System of Chemical Philosophy. John Dalton. Vol. I, part I, 1808; part II, 1810; vol. II, part I, 1827.

Memoirs of the Life and Scientific Researches of John Dalton. Wm. C. Henry. (London, 1854.)

Memoir of John Dalton and the History of the Atomic Theory up to his Time. R. Angus Smith, for the Manchester Literary and Philosophical Society. (London, 1856.)

A New View of the Origin of Dalton's Atomic Theory. Roscoe and Harden. (London, 1896.)

2

BERZELIUS

The dualistic theory

Nothing tends so much to the advancement of knowledge as the application of a new instrument. The native intellectual powers of men in different times, are not so much the causes of the different success of their labours, as the peculiar nature of the means and artificial resources in their possession.... without the Voltaic apparatus there was no possibility of examining the relation of electrical polarities to chemical attractions.

SIR HUMPHREY DAVY, *Collected Works*, edited J. Davy, F.R.S.
(London, 1840), vol. IV 'Elementary Chemical Philosophy', p. 37

Jöns Jacob Berzelius (1779-1848)

An early convert to the Daltonian atomic theory, Berzelius, Professor of Chemistry at the Medico-chirurgical Institute, Stockholm, set himself to establish as firmly as possible its chief experimental foundation—the laws of definite and multiple proportions in chemical compounds. In extending his study widely into the field of non-volatile substances, and especially the metallic compounds, he was compelled to devise and practise novel methods of quantitative analysis, most of which, at least in principle, survive to the present time. For his invention of filter paper and of modern chemical symbols alone he placed all his successors deeply in debt. In assigning formulae to metallic compounds Berzelius frequently supplanted the arbitrariness of Dalton's rules of chemical combination by applications of the law of isomorphism discovered by his pupil and collaborator Mitscherlich (1820)—'the same number of atoms combined in the same manner produce the same crystalline form, which is independent of the chemical nature of the atoms'. In fixing atomic weights he also made use, not without some reservations, of the law of the constancy of atomic heats, published by Dulong and Petit in 1819.

To his genius as an experimenter Berzelius joined a mastery of the exposition of chemistry: in addition to monographs and text-books he produced annually and almost single-handed 'Jahresbericht über die Fortschritte der physischen Wissenschaften' thus founding one of the first, and most enduring, of those 'Annual Reports' which continue to enlighten at the present day.

On 26 June 1800 Sir Joseph Banks, President, notified to the Royal Society the reception from Alessandro Volta (1745–1827), Professor of Natural Philosophy at Pavia, of a letter (published in *Phil. Trans.* of that year) regarding a new means of generating electricity. The original 'Voltaic pile' as it soon came to be called, consisted of a stack of pairs of plates, one of zinc and the other of copper (or better, silver) in contact, each pair being separated from that below it by a piece of fabric soaked in an aqueous solution of common salt. Throughout the pile the same metal, zinc or copper, must be uppermost in all the pairs of plates; so that the terminal plates are of zinc and copper respectively. Volta observed that 'a pile of 40–60 pairs of plates gave sensations resembling those associated with the discharge of a Leyden jar' when bodily contacts were made to wires connected with the terminal plates. Such manifestations were however of short duration, and it was left mainly to Davy, at the Royal Institution, to fashion a serviceable electric battery as a source of steady electric current, as well as to give an explanation of its action.

Davy observed that current flowed between the terminals of a pile only so long as the zinc plates therein were corroded and consumed (with the production of a coating of hydrated zinc oxide) by interaction of the metal with the contiguous aqueous solution. By 1806 he was convinced that 'chemical combination and decomposition are electrical phenomena' and had constructed and used a battery of relatively long life. In this apparatus a plate of zinc and another of copper were separately immersed in each of an extensive series of vessels, or *cells*, filled with an aqueous solution of common alum and nitre (potassium nitrate). The zinc and copper plates in contiguous vessels were connected externally with soldered wires, and current was obtained from the zinc plate at one end of the series and the copper plate at the other. By this arrangement the zinc plates remained clean and the current was not impeded by an insoluble and non-conducting layer of oxide; instead a harmless deposit of alumina accumulated at the base of the cells. The nitrate in the solution was expected to remove the hydrogen necessarily released in the oxidation of the zinc by the aqueous solution. Such agents were afterwards termed 'depolarizers': the Groves cell, which continued as a principal source of laboratory current at least to the close of the nineteenth century, consisted of two vessels, the outer, in which a zinc plate was immersed in dilute sulphuric acid, and an inner, of porous earthenware, which contained a platinum plate immersed in concentrated nitric acid.

During the decade following the publication of Volta's communication, Davy's proposition that chemical action is electrical in nature was supported by the discovery of numerous examples of 'electrolysis', or chemical decomposition by the agency of electric current: by 1808 he had already isolated the metals sodium, potassium, calcium, strontium and barium electrolytically. But even more significant for the understanding of an electrical aspect of chemical action were the observations of Berzelius and his pupils, notably Hisinger, published 1803–1807. In these experiments current was led into aqueous solutions of oxy-salts, for example sodium sulphate, by means of platinum plates (later to be denominated 'electrodes' by Faraday) connected respectively to the zinc and copper plates or *poles* of the battery. By this means it was shown that the bases and acids, from which the salts could be formed by mutual neutralization, were separated, the bases accumulating in the solution in the immediate neighbourhood of the plate connected to the zinc pole and the acids adjacent to the plate attached to the copper pole; in the case of sodium sulphate, soda (regarded as sodium oxide) and sulphuric acid (regarded as sulphur trioxide) were so separated. It appeared to Berzelius that the cause of the separation must lie in the opposite electrifications of the base and acid, and the arbitrary convention was proposed that a base is positively, and an acid negatively, charged, implying that in a battery the zinc plate is the negative and the copper plate the positive pole. It was not until some years later that Berzelius offered what he considered to be an independent proof of the opposed charges on a base and an acid.* Potassium and phosphorus were burnt in dry air beneath two separated plates connected to the opposite poles of a powerful battery. The smoke of potassium oxide was exclusively attracted towards and finally settled upon the negative plate, while that of phosphoric oxide was drawn towards and settled upon the positive plate. Modern theory of the constitution of smokes has compelled a less simple explanation of these phenomena.

Berzelius was further encouraged towards a general electrochemical theory of chemical action from a different quarter. Dalton's views, in agreement with those of most of his contemporaries, on the source of the heat of chemical reaction and particularly of combustion were expressed in the *New System*, I, chap. I, 'On the Quantity of Heat evolved by Combustion'

When certain bodies unite chemically with oxygen, the process is denominated *combustion*, and is generally accompanied with the evolution of heat, in consequence of the diminished (heat) capacities of the products....
...the heat, and probably the light also, evolved by combustion, must be conceived to be derived both from the oxygen and the combustible body; and that each contributes, for aught we know to the contrary, in proportion to its specific heat before combustion.†

It is singular that Dalton, who himself carried out and recorded many measurements of heats of combustion, should have left a quantitative assessment of this principle to Berzelius, who, after showing that it was probably untenable in the combustion of carbon, continues as follows:

Mais choisissons un autre exemple, dont le résultat est encore plus frappant, savoir, la combustion du gaz hydrogène. La chaleur spécifique d'une partie

* *Traité de chimie* (Paris, 1845), I, 104 (authorized translation from the 5th German edition, 1842).
† Pp. 75, 81.

2-2

d'eau est toujours prise pour 1,000; il faut donc que dans cent parties d'eau, il y ait 100,000 de chaleur spécifique....la chaleur spécifique du gaz oxigène est 0,2361; celle du gaz hydrogène, comparée avec celle d'un poids égal d'eau, est 3,2935. Il y a dans 100 parties d'eau 11,1 parties d'hydrogène dont la chaleur spécifique peut être représentée par 36,55, et 88,9 d'oxigène, dont la chaleur spécifique est 20,99. En ajoutant 20,99 à 36,55 on a 57,54 pour la chaleur spécifique du mélange de gaz hydrogène et de gaz oxigène nécessaire pour produire 100 parties d'eau. La combinaison faite, il en résulte de l'eau gazéiforme, dilatée par un feu violent à un volume plusieurs fois plus grand que celui du mélange das deux éléments gazeux. Mais la chaleur spécifique de cette eau refroidie et liquide est 100, c'est-a-dire 42,46 de plus que celle de ses deux éléments a l'état de gaz. D'ou vient donc cette énorme quantité de calorique dégagée par la combustion du gaz hydrogène? Elle n'est point due à un changement de chaleur spécifique, puisqu'il devrait produire un haut degré de froid; ni au dégagement du calorique qui donne la forme gazeuse à l'oxigène et à l'hydrogène, puisque l'eau au moment où elle est formée, produit une vapeur beaucoup plus dilatée que ces éléments gazeuse, et que la condensation de l'eau n'est que l'effet du refroidissement par les corps environnants.*

Since chemical combination (as in a battery) produces electricity, and this in turn causes chemical decomposition (as in electrolysis) it was logical to postulate that the heat (and light) evolved during such typical chemical actions as combustion had an electrical origin, which in Berzelius' view could only reside in the neutralization of opposed electrical charges:

Dans toute combinaison chimique il y a neutralisation des électricités opposées, et cette neutralisation produit le feu, de la même manière qu'elle le produit dans les décharges de la bouteille électrique, de la pile électrique, et du tonnerre....

Si ces corps, qui se sont unis et ont cessé d'être électriques, doivent être séparés...il faut qu'ils recouvrent l'état électrique détruit par la combinaison. Aussi sait-on que, par l'action de la pile galvanique sur un liquide conducteur, les éléments de ce liquide se séparent, que l'oxygène et les acides se rendent du pôle negatif au pôle positif, tandis que les corps combustibles et les bases salifiables sont poussés du pôle positif au négatif.†

Berzelius was fully aware of some of the obstacles confronting the general application of this theory:

* _Théorie_, authorized French edition (Paris, 1835), p. 38. This volume provides a definitive account of Berzelius' ideas. Published originally in Swedish in 1819, it passed through a number of editions in that language and authorized translations in German and French, but no complete English translation appears to have been undertaken. For full title see p. 14, or references, p. 26.

† _Théorie_, p. 46.

(*a*) Certain compounds, such as hydrogen peroxide, chlorine dioxide, and the chloride and iodide of nitrogen, decompose spontaneously with violent explosion in which heat and light are evolved.

(*b*) Concentrated aqueous hydrogen peroxide when mixed with silver oxide is raised to ebullition, and both oxides are decomposed with release of oxygen.

(*c*) If in combination electrical charges are neutralized, what are the forces which, overcome in electrolysis, hold the compound stable?

Such objections were not met until after a century had elapsed, when physicists of the present age fully vindicated Berzelius' opinion that chemical 'synthesis' is an electrical phenomenon, and were able to give an explanation of the energy changes accompanying the exertion of chemical affinity.

The laws of chemical proportions

Like Dr Bostock earlier, but in possession of a much wider range of experimental facts gathered from years of chemical analysis, Berzelius was critical of Dalton's rules. He discussed in detail the methods of arriving at the relative numbers of atoms in compounds, under four heads, of which the following is a summary:

(1) From the relative volumes of combining gases, a source of evidence which Dalton deliberately denied to himself. The dualistic theory necessarily militated against the chemical combination of like atoms, and Berzelius became firmly opposed to Avogadro's conception that the 'ultimate particles' of the elementary permanent gases contained two atoms; but he readily accepted as the basis of Gay-Lussac's laws of gaseous combination a principle that equal volumes of these gases (under the same physical conditions) contained equal numbers of *atoms*. Since it was later proved that the elementary gases concerned have uniformly diatomic molecules, this false assumption did not hinder Berzelius from establishing correct formulae for water, hydrogen chloride, and four oxides of nitrogen. The latter

series he gave as $2N + O$, $N + O$, $2N + 3O$ and $2N + 5O$. For the oxide NO_2, known to Dalton as nitric acid and to his French contemporaries as nitrous acid, Berzelius proposed the composition $3N + 6O$, because it appeared to him to behave chemically as a compound of $N + O$ and $2N + 5O$.

(2) For elements presenting several 'degrees of oxidation', especially metallic elements, the following cases occur of the relative amounts of oxygen combined with a given amount of the element:

Ratio	Probable composition
$1:2$	$R + O$, $R + 2O$ or $R + 2O$, $R + 4O$
$2:3$	$R + O$, $2R + 3O$
$3:4$	$2R + 3O$, $R + 2O$

When for any reason the oxygen compounds do not yield clear indications of relative proportions, compounds of the analogous element, sulphur, may be resorted to.

(3) When a 'base' (*oxide positif*) combines with an 'acid' (*oxide négatif*), the proportions of the oxides are such that the number of oxygen atoms in the acid is an integral multiple of the number in the base, and often this multiple is equal to the number of oxygen atoms in the 'acid'. This rule is exemplified by the salts of nitric, chloric, iodic, sulphuric, and sulphurous acids, and provides a valuable check on the number of oxygen atoms in the 'base'.

(4) When a substance is isomorphous with another in which the number of atoms is known, the number in the first can also be regarded as known, since isomorphism is a consequence of similarity of atomic constitution. In chromic acid and chromic oxide oxygen is in the ratio $2:1$; in chromates the ratio oxygen of base/oxygen of acid is $1/3$; chromic acid and sulphuric acid are isomorphous, as are chromic, ferric, manganic and aluminium oxides. Hence both oxides of chromium contain 3 atoms of oxygen, and since the oxygen ratio is $2:1$ the compositions are $Cr + 3O$, and $2Cr + 3O$.

The dualistic theory

This theory is based on the conception that chemical affinity is a manifestation of electrical attraction, which presupposes that atoms have an electrical constitution.

The atoms of each elementary body have 2 poles, on which the two opposite electricities are accumulated in different proportions, according to the nature of the bodies: for example, oxygen has a large negative quantity, and a small positive quantity, while potassium has the opposite disposition: hydrogen has nearly equal quantities of each electricity.*

Both electricities must be postulated to be present in every atom, to account for such facts as that sulphur combines vigorously both with electro-positive potassium and electronegative oxygen.

On the basis of this principle and the general chemical behaviour of their known oxygen compounds, Berzelius drew up a provisional electrochemical series of the elements:

Predominance of negative O, S, N, F, Cl, Br, I, P, As, Cr, V, B, C, Sb,
pole decreasing Te, Ti, Si, (H)

Predominance of positive (H), Au, Pt, Hg, Ag, Cu, Bi, Sn, Pb, Cd, Co,
pole increasing Ni, Fe, Zn, Mn, Al, Mg, Ca, Sr, Ba, Li, Na, K

Les corps simples électropositifs ainsi que leurs oxides, prennent toujours l'électricité positive lorsqu'il rencontrent des corps simples ou des oxides appartenant aux électronégatifs. Les oxides des électropositifs se comportent toujours avec les oxides des électronégatifs comme les bases salifiables avec les acides.†

It had been observed by Davy and others that when, in one of the pairs of metals in a cell of a battery, the metals are interchanged in position and the battery circuit is closed, then in the reversed cell the copper (or silver) plate is corroded and dissolved, while the zinc plate remains intact: interchanging positions apparently entailed also an interchange of chemical properties. This result harmonized with Berzelius' series of metals, for the effect of the battery on the reversed cell is to force a positive charge into the copper plate and an opposite charge

---

* Gmelin's *Handbuch der Chemie*, translation by Watts (Cavendish Society, London, 1848).
† *Théorie*, p. 48.

23

into the zinc plate, thus producing an increase of electro-positive polarity in the former and an increase of electro-negative polarity in the latter: the copper is moved in the series towards potassium, and the zinc towards gold.

It will be seen that in the second part of the series, following hydrogen, the order of the metals is strikingly similar to that of the modern electrochemical series of the metals, which is based firmly on well-established numerical data.

Berzelius' theory of salt formation may be conveniently symbolized as below, the magnitude of the polar charges of electricity being broadly suggested by different sizes of the signs $+$ and $-$; R' and R'' are 'elementary radicals':

$$\left.\begin{array}{l} -R' + \\ +O - \end{array}\right\} = (R'O) +, \text{ base;} \qquad \left.\begin{array}{l} +R'' - \\ -O + \end{array}\right\} = (R''O) -, \text{ acid.}$$

Bases and acids were *primary compounds*;

$$\left.\begin{array}{l} (R'O) + \\ (R''O) - \end{array}\right\} = (\text{salt}) \; {}^{+}_{-}, \quad \text{a } secondary \; compound;$$

$$\left.\begin{array}{l} {}^{+}_{-} (\text{salt}) \\ (\text{water}) \end{array}\right\} = \text{salt hydrate,} \quad \text{a } tertiary \; compound.$$

The whole process of the production of (hydrated) salt was regarded as a staged neutralization of electrical charges, which became progressively smaller in absolute magnitude; tertiary compounds, such as salt hydrates, were therefore thermally unstable. The most positive metals (e.g. sodium or potassium) yielded the most stable bases, which were not decomposed in the aqueous electrolysis of their salts: less stable bases from less positive metals (copper, silver) separated in the electrolysis of their salts, and underwent further electrolytic decomposition, with deposition of the metal.

The base of the ammonium salts, regarded as $(NH_4)_2O$, contained two electro-negative elements and was therefore exceptionally weak, that is, only feebly positive: ammonium salts, like tertiary compounds, were consequently thermally unstable.

The dualistic theory involved the proposition that all chemical compounds should be divisible into two parts with opposite net electrical charges, an idea that could be pursued with some success in inorganic chemistry, but was ill adapted to the organic branch of the science. Dumas discovered, in the period 1834–1838, the phenomenon he termed 'metalepsy': in organic compounds, for example acetic acid, negative chlorine may replace hydrogen atom for atom without essential change in the chemical nature of the compound, and acetic acid can be regenerated from the chlorinated acid. On this and other facts of organic chemistry unamenable to the dualistic theory Berzelius tersely commented 'Les phénomènes électriques se manifestent principalement dans la nature inorganique: dans la nature organique, l'état des choses est différent'.* Although no means existed for assessing the actual amounts of the polarized charges it was necessary to take into account two aspects of the atomic polarities: the ratio of the charges as well as their absolute magnitudes. Moreover to explain the marked effect of temperature on chemical affinity it had to be assumed that all these conjectural quantities were functions of the temperature. The flexibility introduced by the possibility of such varied adjustments made it easy to apply the theory widely, without arousing conviction that the truth had been reached.

In his exposition of the dualistic theory in the fifth and last edition of his text-book of chemistry† Berzelius introduced a new postulate, which appeared to give a partial answer to questions raised in his earlier writing:

Si les atomes n'avaient qu'un seul axe de polarisation, ils ne pourraient se combiner que dans le rapport d'un atome d'un élément avec un atome de l'autre; et, dans ce cas, leurs pôles électriques opposés deviendraient leurs points de combinaison. Or, il n'en est pas ainsi. Nous avons déjà vu qu'un corps élémentaire peut se combiner avec un, deux, trois et plus d'atomes d'un autre corps; il suit de là que les atomes possèdent, dès le principe, plusieurs axes de polarisation, ou qu'ils les acquièrent sous l'influence de certaines circonstances; il se forme donc alors plusieurs pôles, dont chacun correspond à un pôle opposé des atomes multiples combinés avec lui. Il arrive ordinairement que les corps élémentaires dans lesquels l'électricité positive est prédominante, se combinent de préférence avec plusieurs atomes de corps dans lesquels l'électricité négative prédomine; . . . il paraît

---

* *Théorie*, p. 61.
† German edition 1842, French edition 1845.

suivre de là, que l'électricité positive a plus de tendance à se partager en plusieurs points pôlaires que l'électricité négative. On ne sait pas positivement si, dans ce cas, plusieurs axes de polarisation se coupent et forment plusieurs pôles négatifs libres, ou s'ils coincident en un seul point, c'est-a-dire, s'ils forment un pôle commun; tout ce que nous savons par l'expérience, c'est qu'un corps électro-positif, qui se combine avec 1 atome d'un corps électronégatif, peut encore être fortement électro-positif; qu'avec 2 atomes il l'est peu; ou il devient tout à fait indifférent, et qu'avec un plus grand nombre d'atomes il devient de plus en plus électro-négatif, ce qui prouve que l'intensité du pôle négatif s'accroît considérablement, et l'opinion que la polarité est accumulée dans un seul point se trouve ainsi justifiée.*

In the earlier account of the theory, published in 1835, we find:

Quand même il serait suffisament prouvé que les corps...sont composés d'atomes indivisibles, il ne s'ensuivrait pas que les phénomènes des proportions chimiques...doivent necessairement avoir lieu. Il faut encore l'existence de certaines lois qui règlent les combinaisons des atomes et qui leur assignent de certaines limites....C'est donc principalement de ces lois que dépendent les proportions chimiques....†

Nous ignorons la cause des limites assignées aux combinaisons des atomes entre eux, et nous ne pouvons même former à ce sujet aucune conjecture admissible.‡

It is difficult to disbelieve that Berzelius was reaching towards an idea of specific valency, which is essentially concerned with 'assigning certain limits', both upper and *lower*, to the combination of atoms. Dalton could give what might then appear a valid reason for an upper limit (p. 8), but a lower limit greater than unity would have seemed to him impossible. If in an atom possessing 'multiple axes of polarization' terminating in a common central charge the (arbitrary) signs of the electricities are interchanged, the constitution of the combining atom presented could be seen as singularly prophetic of that accepted today.

BIOGRAPHICAL AND OTHER REFERENCES

Berzelius' Werden und Wachsen (1779–1821). H. G. Söderbaum. (Leipzig, 1899.)

La théorie des proportions chimiques et de l'influence de l'électricité dans la nature inorganique. Authorized French edition. (Paris, 1835.)

Traité de chimie, vol. 1. Authorized translation from the 5th German edition, 1842. (Paris, 1845.)

* French edition (1845), p. 105.   † *Théorie*, p. 18.
‡ *Théorie*, p. 25.

3

FRANKLAND, COUPER AND KEKULE

Valency as an explicit doctrine

Before leaving the subject of the atom it is desirable to refer to the new idea which is revolutionizing chemistry, namely, that the different atoms have different exchangeable values, or *valencies*. An atom is the seat of attractive forces, and the valency is the number of centres of such force.
From an Essay by 'Sigma' in the *English Mechanic*, vol. x, issue of
19 November 1869: according to O.E.D. the first use
in print of the term 'valency'

1. Edward Frankland (1825-1899)

During the years when Frankland was developing his theory of valency (1852–1866) two other portentous events heralded the advent of modern chemistry: the promulgation in 1857 by Clausius of the kinetic theory of the gaseous state, and the acceptance of Avogadro's principle at that 'general council' of chemists held at Karlsruhe in 1860. In fact the physicist preceded the chemists in that matter, for he relied on the truth of Avogadro's principle as an essential basis of the exposition of the kinetic theory.* Before 1860 (and in some quarters for a generation later) the utmost confusion prevailed in the assignment of atomic weights, and not uncommonly the system adopted in publications depended on the laboratory in which the work originated.

Whether an atom of water contains the same quantity or double the quantity of hydrogen that is contained in an atom of hydrochloric acid and whether the atomic weight of oxygen is 8 or 16 are concrete examples of the

* *Phil. Mag.* (4), **14**, 108 (1857).

27

many disputed questions which lie at the very basis of scientific chemistry. ...It is evidently most inconsequent to interpret a certain set of facts in one special manner and a precisely similar set of facts in an opposite manner.... The majority of English chemists represent the atomic weight of carbon by 6, that of oxygen by 8, and that of sulphur by 16. Dr. Frankland would double the atomic weight of carbon, but would retain the old atomic weights of oxygen and sulphur. Mr. Griffin, who lays claim to priority in doubling the atomic weights of carbon and oxygen ridicules the notion of doubling that of sulphur. Dr. Williamson, Mr. Brodie and myself have for a long time advocated the doubling of all three.

For silicon Thomson took 7·12, assuming the chloride to be SiCl: Gmelin prefers $SiCl_2$ and 14·25, and Berzelius 21·37 and $SiCl_3$; whereas in my opinion the balance of argument is in favour of $SiCl_4$ and 28·50.*

Gmelin's system of 'equivalents' had gained some ascendancy over others, and included the following (correct values in parentheses):

$$H = 1, C = 6 \ (12), N = 14, O = 8 \ (16), Cl = 35,$$

$$P = 31, S = 16 \ (32), I = 127, As = 75,$$

$$Sb = 122, Zn = 32 \cdot 5 \ (65), Hg = 200, Sn = 59 \ (119).$$

We owe solely to Cannizzaro that new definition of atomic weight which alone brought a final agreed settlement: 'the atomic weight of an element is the least weight of it present in the molecular weight of any of its compounds'. In 1858 Avogadro's devoted compatriot had published in the journal *Il Nuovo Cimento* an essay, under the title 'An Outline of a Course of Chemical Philosophy', which contained the first statement of this definition. Copies of the essay were circulated among the members of the Karlsruhe Congress. Its influence upon at least one chemist was dramatically expressed by Lothar Meyer: 'I read it again and again, and was astonished at the light thrown on the most important points at issue. It was as though the scales fell from my eyes, doubt vanished, and was replaced by a feeling of peaceful certainty.' Meyer effectively supported Cannizzaro's mission in his widely read book *Die modernen Theorien der Chemie* which appeared in 1864. Conver-

* Odling, 'The Atomic Weights of Oxygen and Water', *Quart. J. chem. Soc.* **11**, 107 (1859).

sion did not come so quickly in all quarters, for Mendeleef, who was present at the Congress, could write as late as 1889 'Is it so long since many refused to accept the generalizations involved in the law of Avogadro and Ampère? We still may hear the voices of its opponents: they enjoy perfect freedom, but vainly will their voices rise so long as they do not use the language of demonstrated facts.'*

Frankland's early career was closely connected with the first foundation in England of institutions specially devoted originally to the teaching of chemical science. Liebig's visit to England in 1842 had been largely occupied with public lectures intended to arouse interest in the application of chemistry to agriculture, and as a direct result, in 1845, the Royal College of Chemistry was opened in Oxford Street, London, by a private committee in which the Prince Consort played a prominent part. Upon Liebig's recommendation A. W. Hofmann was chosen as Professor of Chemistry. In the terms of his office the position of applied chemistry was specially safeguarded in research and teaching. Four years later, on the foundation, by a similar committee, of Owen's College in Manchester, Frankland, recently returned from Germany where in a brilliant career he had worked with Bunsen and Liebig, was, at the early age of 24, appointed Professor of Chemistry.

In 1857 Frankland returned to London, where as a youth he had assisted in the work of Lyon Playfair's laboratory and, after a short period as Faraday's successor at the Royal Institution, he succeeded Hofmann in 1865. When he retired twenty years later, the original Royal College of Chemistry, transferred to South Kensington in 1872, was about to be absorbed into the Chemical Department of the Royal College of Science.

It must be remembered that most of the organo-metallic compounds, on the constitution of which Frankland based his concept of valency in the period 1852–1866, are insufficiently volatile for determinations of their vapour densities, and it was not until 1886 that Raoult first announced the principle of the cryoscopic method of fixing molecular weights. Frankland was

* Faraday Lecture of the Chemical Society, 1889.

therefore compelled to rely very largely upon gravimetric analyses, interpreted by Gmelin's 'equivalents' (p. 28), as the means of arriving at chemical constitution. It was a singularly auspicious circumstance that the inconsistencies in the Gmelin system of 'atomic weights' should so often result in constitutional formulae displaying at a glance a valency principle. Thus the 'triatomicity' of arsenic was immediately obvious when arsenious oxide was expressed as

$$\text{As} \begin{cases} \text{O} \\ \text{O}, \\ \text{O} \end{cases}$$

but it is more than doubtful whether Frankland would have ventured such an opinion when confronted by the present-day formula As_4O_6.

The liquid compound, later to be named *cacodyl oxide* by Berzelius, was discovered in 1760 by Cadet among the volatile products of the dry distillation of a mixture of arsenious oxide and potassium acetate. In the period 1837–1843 Bunsen led in his laboratory a full examination of this compound and its reactions,* and instigated an investigation of similar compounds of antimony. Work in the field of organo-metallic bodies was still proceeding when Frankland entered Bunsen's laboratory, and undoubtedly inspired his researches after his return to England.

In a publication in 1852† Frankland described his recent preparation of organo-metallic compounds of zinc, tin and mercury, and drew attention to the relations between the following formulae (modern formulae are shown on the right of rule):

$\text{As} \begin{cases} \text{S} \\ \text{S} \end{cases}$ $\text{As} \begin{cases} \text{C}_2\text{H}_3 \\ \text{C}_2\text{H}_3 \end{cases}$	As_4S_4, realgar	
	$As_2(CH_3)_4$, cacodyl	
$\text{As} \begin{cases} \text{O} \\ \text{O} \\ \text{O} \end{cases}$ $\text{As} \begin{cases} \text{C}_2\text{H}_3 \\ \text{C}_2\text{H}_3 \\ \text{O} \end{cases}$	$[As(CH_3)_2]_2 . O$, cacodyl oxide	

* *Liebigs Ann.* **37**, 1 (1841); **42**, 14 (1842); **46**, 1 (1843).
† *Phil. Trans.* **142**, 417 (1852).

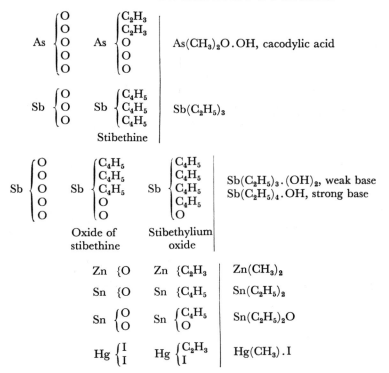

When the formulae of inorganic compounds are considered even a super-ficial observer is struck with the general symmetry of their construction; the compounds of nitrogen, phosphorus, arsenic and antimony especially exhibit the tendency of these elements to form compounds containing 3 or 5 equivalents of other elements, and it is in these proportions that their affinities are best satisfied. In the *ternal* group we have

$$NO_3, NH_3, PO_3, PH_3, PCl_3, SbO_3, SbH_3, SbCl_3, AsO_3, AsH_3, AsCl_3, \text{ etc.,}$$

and in the five-atom group

$$NO_5, NH_4O, NH_4I, PO_5, PH_4I, \text{ etc.}$$

Without offering any hypothesis regarding the cause of this symmetrical grouping of atoms, *it is sufficiently evident, from the examples given, that such a tendency or law prevails, and that, no matter what the character of the uniting atoms may be, the combining power of the attracting element,* if I may be allowed the term, *is always satisfied by the same number of these atoms.*

Stibethine furnishes us with an example of the operation of the law of symmetrical combination and shows that the formation of a five-atom group from one containing three atoms can be effected by the assimilation of two atoms, *either of the same or of opposite electrochemical character.* Is this behaviour common also to compounds of nitrogen, phosphorus and arsenic: and can the position of each of the 5 atoms with which these elements combine be indifferently occupied by an electronegative or an electropositive element?*

Two points emerge: some reluctance to oppose the dualistic theory, and the early recognition that elements may exhibit more than one 'atomicity'.

A communication three years later† is mainly concerned with the preparation and reactions of zinc ethide (zinc diethyl). This compound inflames on contact with air, and Frankland found the solid product to be mainly zinc oxide, mixed with some zinc acetate, the production of which is ascribed to the combustion of the expelled radical, giving acetic acid which attacks a part of the oxide. The action of water, halogens and sulphur upon zinc ethide was expressed in the following equations:

$$\left.\begin{array}{l}ZnC_4H_5\\HO\end{array}\right\} = \left\{\begin{array}{l}ZnO\\C_4H_5.H\end{array}\right. \qquad \left.\begin{array}{l}ZnC_4H_5\\2I\end{array}\right\} = \left\{\begin{array}{l}ZnI\\C_4H_5I\end{array}\right.$$

$$\text{(and similarly for bromine)}$$

$$\left.\begin{array}{l}ZnC_4H_5\\2S\end{array}\right\} = \left\{\begin{array}{l}ZnS\\C_4H_5S\end{array}\right.$$

In the inorganic combinations of zinc the metal unites with one atom only of other elements. The atom of zinc appears therefore to have only one point of attraction, and hence *notwithstanding the intense affinities of its compound with ethyl* any union with a second body is necessarily attended by expulsion of the ethyl.

In delivering the Bakerian Lecture to the Royal Society in 1859 Frankland gave the following equations expressing his researches on the organo-metallic compounds of tin:

$$(1) \quad \left.\begin{array}{l}C_4H_5I\\Sn\end{array}\right\} = \left\{Sn\left\{\begin{array}{l}C_4H_5\\I\end{array}\right.\right. \qquad Sn\left\{\begin{array}{l}C_4H_5\\I\end{array}\right.\left.\right\} = \left\{\begin{array}{l}SnC_4H_5\\ZnI\end{array}\right.$$

* The term 'alkyl radical' reflects the current opinion that such groups possessed an electro-positive character, indicated, for example, by the difference between cacodylic acid and cacodyl oxide, which is basic (formulae on previous page).

† *Phil. Trans.* **145**, 259 (1855).

$$(2) \quad \left. \begin{array}{l} SnC_4H_5 \\ O \end{array} \right\} = \left\{ Sn \begin{array}{l} C_4H_5 \\ O \end{array} \right. \qquad \left. \begin{array}{l} SnC_4H_5 \\ I \end{array} \right\} = \left\{ Sn \begin{array}{l} C_4H_5 \\ I \end{array} \right.$$

$$(3a) \quad 2SnC_4H_5 \} = \left\{ \begin{array}{l} Sn \begin{array}{l} C_4H_5 \\ C_4H_5 \end{array} \\ Sn \end{array} \right. \qquad (3b) \quad \left. \begin{array}{l} Sn \begin{array}{l} C_4H_5 \\ I \end{array} \\ ZnC_4H_5 \end{array} \right\} = \left\{ \begin{array}{l} Sn \begin{array}{l} C_4H_5 \\ C_4H_5 \end{array} \\ ZnI \end{array} \right.$$

(distillation)

$$(4) \qquad \left. \begin{array}{l} 2Sn \left\{ \begin{array}{l} C_4H_5 \\ C_4H_5 \end{array} \right. \\ HCl \end{array} \right\} = \left\{ \begin{array}{l} Sn_2 \left\{ \begin{array}{l} (C_4H_5)_3 \\ Cl \end{array} \right. \\ C_4H_5.H \end{array} \right.$$

Equations (2) are to be contrasted with the different behaviour of zinc ethide (p. 32), and equation (3a) recalls the action of hot aqueous alkali upon stannous oxide. In regard to reaction (4) Frankland writes

It is important to ascertain the deportment of stannic ethide with negative elements, since, if it were found to be capable of direct combination, its analogy to inorganic stannic compounds (for example, stannic oxide or iodide) would be to a great extent destroyed. Like zinc ethide, however, stannic ethide is incapable of combining with halogens without the expulsion of at least an equivalent of the ethyl it contains.

The lecture concludes with the following reactions confirming that mercury reaches a saturated condition as a dyad:

$$(1) \quad \left. \begin{array}{l} C_2H_3I \\ Hg \end{array} \right\} = \left\{ Hg \begin{array}{l} C_2H_3 \\ I \end{array} \right. \qquad \left. \begin{array}{l} Hg \left\{ \begin{array}{l} C_2H_3 \\ I \end{array} \right. \\ 2ZnC_4H_5 \end{array} \right\} = \left| \begin{array}{l} Hg \left\{ \begin{array}{l} C_4H_5 \\ C_4H_5 \end{array} \right. \\ ZnC_2H_3 \\ ZnI \end{array} \right.$$

$$(2a) \qquad \left. \begin{array}{l} Hg \left\{ \begin{array}{l} C_4H_5 \\ C_4H_5 \end{array} \right. \\ HgCl_2 \end{array} \right\} = \left\{ 2Hg \begin{array}{l} C_4H_5 \\ Cl \end{array} \right.$$

$$(2b) \qquad \left. \begin{array}{l} Hg \left\{ \begin{array}{l} C_4H_5 \\ Cl \end{array} \right. \\ ZnC_2H_3 \end{array} \right\} = \left\{ 2Hg \begin{array}{l} C_4H_5 \\ C_2H_3 \end{array} \right. \\ ZnCl$$

In a long essay of some sixty printed pages, entitled rather unrealistically 'A Discourse to the Members of the Chemical Society of London'* Frankland renders a detailed account of

* *Quart. J. chem. Soc.* **13**, 177–235 (1861). Cited hereafter as 'Discourse'.

the then known organo-metallic compounds, and sums up their significance as follows:

The behaviour of the organo-metallic bodies teaches a doctrine which affects chemical compounds in general, and which may be called the doctrine of atomic saturation; each element is capable of combining with a certain number of atoms; and this number can never be exceeded, although the energy of its affinities may have been increased up to this point.

At the close of the essay he indicates how his theory of 'atom-icity' could be extended to (aliphatic) organic compounds in general, without revealing any cognizance that Kekulé had made similar advances three years previously.*

The time of Frankland's publication lay in the critical years between the long domination of Gmelin's 'equivalents' with other arbitrary systems of 'atomic weights', and the acceptance of the true values based upon Cannizzaro's principle (p. 28): and, to bring into the theory of 'atomicity' the uniformity which he was convinced existed, he was obliged to resort to the device of the 'double atom'. For example, in his Bakerian Lecture (1859) he gave for four stannic compounds the formulae

$$\text{Sn} \begin{cases} C_4H_5 \\ C_4H_5 \end{cases}, \quad \text{Sn} \begin{cases} C_4H_5 \\ I \end{cases}, \quad \text{Sn} \begin{cases} C_4H_5 \\ O \end{cases}, \quad \text{Sn}_2 \begin{cases} (C_4H_5)_3 \\ Cl \end{cases},$$

which were the simplest possible on the basis $Sn = 59$, $C = 6$, $O = 8$. A principle of uniform atomicity is secured only by assigning 'quadratomicity' to the 'double atom' Sn_2 and writing such formulae as

$$\text{Sn}_2 \begin{cases} C_4H_5 \\ C_4H_5 \\ C_4H_5 \\ C_4H_5 \end{cases} \quad \text{Sn}_2 \begin{cases} C_4H_5 \\ C_4H_5 \\ C_4H_5 \\ Cl \end{cases}.$$

Similarly it appeared necessary to write for methane and its substitution products, not the simplest possible formulae,

$$CH_2, \quad C_2H_3Cl, \quad CHCl, \quad C_2HCl_3, \quad CCl_2,$$

* *Liebigs Ann.* **106**, 129 (1858): see p. 53.

but rather

$$C_2 \begin{cases} H \\ H \\ H \\ H \end{cases} \quad C_2 \begin{cases} H \\ H \\ H \\ Cl \end{cases} \quad C_2 \begin{cases} H \\ H \\ Cl \\ Cl \end{cases}, \quad C_2 \begin{cases} H \\ Cl \\ Cl \\ Cl \end{cases}, \quad C_2 \begin{cases} Cl \\ Cl \\ Cl \\ Cl \end{cases},$$

and 'quadratomicity' was accorded to the 'double atom' C_2.

It is difficult to believe that Frankland did not entertain the clear implication that the atomic weights of tin, carbon and many other elements were assigned at half their true values, but in proposing corrections he may well have feared to arouse again a spate of recriminations such as had disfigured chemistry in former years. Such a view is the more probable from the circumstance that in his last communication to the Chemical Society on the subject of 'atomicity', in 1866, he adopted the 'reformed' atomic weights without comment.

In the 'Discourse'* Frankland reiterates his views on 'polyatomicity':

...in the combinations of polyatomic metals we frequently notice from the lowest to the highest compound one or more intermediate points of exalted stability; thus antimony has a teratomic stage of comparative stability; nitrogen, phosphorus, and arsenic whilst exhibiting a similar teratomic stage, have also a biatomic one, though of greatly inferior stability; whilst the existence of protoxide of nitrogen renders it more than probable that nitrogen has a third and uniatomic stage.

In bodies possessing at least one stage of stability below saturation and in which all the atoms united with the polyatomic element are of the same kind, the stage of maximum stability is very rarely that of saturation. Thus in nitrogen, arsenic and bismuth the stage of maximum stability is decidedly the teratomic one, whilst in phosphorus compounds the points of maximum stability and of saturation generally coincide. Nitrogen exhibits a much stronger tendency towards universal teratomic stability than its chemical associates.

The atomicity of carbon is treated as follows:

The double atom of carbon, like that of tin, is quadratomic in perchloride of carbon and in carbonic acid—

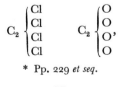

$$C_2 \begin{cases} Cl \\ Cl \\ Cl \\ Cl \end{cases} \quad C_2 \begin{cases} O \\ O \\ O \\ O \end{cases},$$

* Pp. 229 *et seq.*

and biatomic in protochloride of carbon, allyl iodide and carbonic oxide—

$$C_2 \begin{cases} Cl \\ Cl \end{cases} \qquad C_2 \begin{cases} C_4H_5 \\ I \end{cases} \qquad C_2 \begin{cases} O \\ O \end{cases}.$$

'Protochloride of carbon' (perchloro-ethylene) was at this time obtained by saturating 'olefiant gas' (ethylene) with chlorine and submitting the product (perchloro-ethane) to pyrolysis at red heat. It is curious that Frankland does not comment on the fact that the 'protochloride' could not be obtained from 'perchloride of carbon' nor is it convertible into that compound by the action of chlorine. It is also evident that he hesitated at this stage to propose a formula for ethylene.

In 1866 Frankland communicated to the Chemical Society 'Contributions to the Notation of Organic and Inorganic Compounds', and published 'Lecture Notes for Chemical Students'. In these works, with which he was to take his leave of the 'doctrine of atomicity', at least in the matter of publication he finally ushered the doctrine into modern chemistry, by adopting true atomic weights ($C = 12$, $O = 16$, etc.), and, in the 'Lecture Notes', the notation of 'valency bonds' upon which structural formulae continue to be based.

The following extracts, slightly abridged, are from the paper to the Chemical Society:*

As examples of the mode of notation the following may be adduced—

$$C^{IV} \begin{cases} CH_3 \\ H \\ H \\ OH \end{cases} \qquad C^{IV} \begin{cases} CH_3 \\ H \\ H \\ O\ C_2H_5 \end{cases} \qquad C^{IV} \begin{cases} CH_3 \\ O \\ OH \end{cases} \qquad C^{IV} \begin{cases} CH_3 \\ O \\ H \end{cases} \qquad C^{IV} \begin{cases} CH_3 \\ O \\ CH_3 \end{cases}$$

Ethylic alcohol	Ethylic ether	Acetic acid	Acetic aldehyde	Acetone

The meaning of these formulae is obvious; that of ethylic alcohol, for instance, represents an atom of tetrad carbon united to the carbon of methyl, to each of two atoms of hydrogen, and to oxygen, the latter dyad element being also united to an atom of hydrogen. This is a true representation of the internal arrangement of the atoms composing alcohol, not, indeed, of their relative positions in space, but of the mode in which they are held together.... Such a formula does not deny the exercise of a certain influence by each atom over every other atom in the compound, but it does not allow the conception of an exchange of the atom of hydrogen, for instance, in

* *J. chem. Soc.* **19**, 372 (1866).

union with oxygen for one in union with carbon. On no other hypothesis can the fact of isomerism receive an intelligent explanation.

In writing the formula of any compound, the element possessing the highest atomicity is placed first, that is to the left, and this element is considered to be directly united with all the *active bonds* of the other elements following upon the same line; thus the formula of sulphuric acid $SO_2(OH)_2$ signifies that the hexad atom of sulphur is combined with the four bonds of the two atoms of oxygen, and also with the two bonds of two atoms of hydroxyl....By the term *bond* I intend merely to give a more concrete expression to such names as 'an atomicity', 'an atomic power', or 'an equivalence'. A monad is an element having one bond, a dyad two bonds, etc. By this term I do not intend to convey the idea of any *material* connection, the bonds actually holding together the atoms of a compound being probably much more like those which connect the members of the solar system.

The number of bonds possessed by an element, or its atomicity, is, apparently at least, not a fixed and invariable quantity; thus nitrogen is equivalent to five atoms of hydrogen in ammonium chloride, NH_4Cl, to three in ammonia, NH_3, and to one in nitrous oxide, ON_2. But it is found that this variation always takes place by the disappearance or development of an even number of bonds; thus nitrogen is a pentad, a triad, or a monad; phosphorus and arsenic, either pentads or triads; carbon and tin, tetrads or dyads; and sulphur, selenium, and tellurium either hexads, tetrads or dyads. These facts can be explained by a very simple assumption, viz., that *one or more pairs of bonds, having saturated each other, become*, as it were, latent. The maximum number of bonds of an element may be termed its *absolute atomicity*, those united together its *latent atomicity*, and the number actually engaged in a compound its *active atomicity*.

The exceptions to the proposed rule apparent in the current formulae $HgCl$, $HgCl_2$ and $FeCl_2$, $FeCl_3$ were met by observing that 'the vapour density of ferric chloride shows that its formula must be doubled, and there are strong reasons for doubling the formula of calomel'.

Although, as the above extract shows, Frankland still held to the view that carbon could function as a dyad or tetrad, he now committed himself to a formula for ethylene, in the brief statement:

When the elements connected by a bracket are united *by more than one bond*,* this circumstance is indicated by atomicity marks to the left of the bracket—

$$\begin{cases} CH_3 \\ CH_3 \end{cases} \qquad {}_{''} \begin{cases} CH_2 \\ CH_2 \end{cases}$$

Ethane   Ethylene

* Present author's italics.

The formula concedes 'quadratomic' carbon in ethylene, but Frankland did not at this point reconsider the formulae for perchloroethylene and allyl iodide he had previously put forward (p. 36).

Later in the paper he applied his conception of the two possible atomicities of the carbon atom in a tentative explanation of the isomerism of maleic and fumaric acids, the formulae of which he suggested must ultimately be chosen from among such expressions as

$$_{''}\begin{cases} CH(COHO) \\ CH(COHO) \end{cases} \qquad \begin{cases} CH_2(COHO) \\ {''}C(COHO) \end{cases} \qquad \begin{cases} COHO \\ C({''}CH)H. \\ COHO \end{cases}$$

'The symbol $''C$ signifies carbon with two of its bonds latent, as in carbonic oxide, $''CO$.' This problem, with others of a similar nature, was not solved until, in the years 1887–1889, Wislicenus applied to them the recently recognized vectorial properties of carbon bonds.

In the preface to 'Lecture Notes' (dated 15 September 1866), which is notable for including one of the few references he made to his great contemporary Kekulé, Frankland writes

The contents...are to great extent a transcript of the notes of my lectures delivered at the Royal College of Chemistry during the Winter Session of 1865–66.

To illustrate important constitutional formulae I have extensively adopted the graphic notation of Crum Brown, which appeared to me to possess important advantages over that first proposed by Kekulé....I am aware that graphic formulae may be objected to, on the ground that students...will be liable to regard them as representations of the actual physical position of the atoms of compounds. I have not found this evil to arise, and even if it did I should deprecate it less than ignorance of all notion of atomic constitution.

That there should appear, in his paper to the Chemical Society of the same year, no hint of his favourable regard for graphic formulae is due perhaps to Frankland's political sense of the new storm they were to arouse, reaching a climax in the invectives of the redoubtable Kolbe in the pages of the *Journal für praktische Chemie*, of which he was editor. Kolbe and Frankland were friends and contemporaries in their student days in Germany, and had conducted a lengthy correspondence, at the close of which Kolbe publicly confessed that he had been won over, albeit reluctantly, to his friend's early views on atomicity.* Frankland's preface, quoted above, shows that he was not to be fully absolved from the current agnosticism about the physical significance of any type of graphic formula, but when in 1875 Le Bel and van't Hoff undermined the basis for such an attitude Kolbe's malicious stupidity reached limits fortunately rare, if not unique, in the annals of chemistry.†

It is interesting to find a contrary opinion of chemical formulae expressed in 1852, the year of Frankland's first paper, by Williamson, with whom Kekulé regularly exchanged views while working in London: 'Formulae may be used as an actual image of what we rationally suppose to be the arrangement of the constituent atoms in a compound, as an orrery is an image of what we conclude to be the arrangement of our planetary system.' Unfortunately Williamson did not see fit to press such views on his contemporaries.‡

The graphic formula with drawn valency links is so natural and obvious a manner of expressing the regimen of valency in a chemical compound that it could hardly fail to be suggested by more than one chemist. Two young assistants to Playfair at Edinburgh, Couper and Crum Brown, appear to have jointly sponsored a notation which in essentials survives to the present day. Owing to the early failure of Couper's health (p. 46) it fell to Crum Brown to develop their ideas.§ In his earliest formulae the constituent atoms, represented as labelled circles, were given the number of radiating strokes or 'bonds' warranted by their respective valencies, and then assembled into a formula so that every such 'bond' became set in line with, but not joined

* *Liebigs Ann.* **101**, 257 (1857).
† See *Proc. chem. Soc.* 1959, 285.
‡ 'On the Constitution of Salts', *Quart. J. chem. Soc.* **4**, 351 (1852).
§ *Proc. roy. Soc. Edinb.* **5**, 429 (1865); 561 (1866); *Trans. roy. Soc. Edinb.* **23**, 707 (1864).

to, one in another atom. For example,

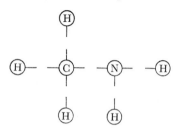

A formula was valid only if all 'bonds' were paired. The further step of joining each pair of 'bonds' to form interatomic links was more than a mere convenience, for it implied a process of mutual valency saturation, not to be fully understood until the present century.*

In 'Lecture Notes' the following formulae are given for elementary molecules:

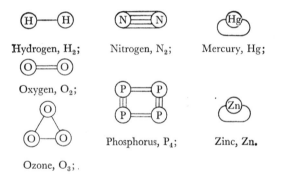

'An element with an even number of bonds (as in mercury and zinc) can exist as a monatomic molecule, its own bonds satisfying each other', a statement notable for the use of 'monatomic' in its modern sense. The supposed 5-membered link in nitrogen reveals almost luridly how far the current agnostic attitude to graphic formulae suppressed any speculation about the possible directional properties of 'bonds'. Of the numerous formulae provided for inorganic compounds the following are typical:

* See further on Couper's work, p. 49.

Carbon dioxide, CO_2;

Sulphur trioxide, SO_3;

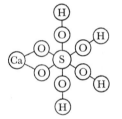

Nitric acid, HNO_3;

Ammonium chloride, NH_4Cl;

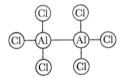

Calcium sulphate dihydrate,
$CaSO_4.2H_2O$;

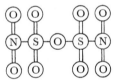

Calcium sulphate,
$CaSO_4$;

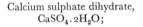

Aluminium chloride, Al_2Cl_6;

'Chamber crystals'
(sulphuric acid process).

Among the formulae ascribed to organic compounds the following appear:

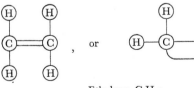

or

Ethylene, C_2H_4;

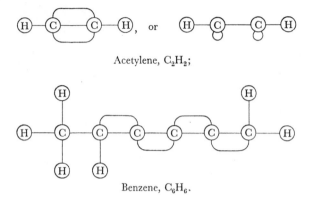

Acetylene, C_2H_2;

Benzene, C_6H_6.

Only in one instance (on p. 202) does Frankland give ethylene the first type of formula above, in which he represents doubly linked carbon atoms as he invariably does oxygen, nitrogen, sulphur, etc., in a similar state. The preference for a more expanded form of the multiple carbon link is nowhere explained. In the alternative formula for acetylene two bonds in each carbon are shown as 'self-neutralizing' (p. 37). In 1868 Frankland's interests became wholly diverted to applied chemistry, by his appointment to a Royal Commission of three members 'to inquire into the pollution of rivers and the domestic water supplies of Great Britain'. From this time, at least in the matter of new publications, he held himself completely aloof from the later development of the chemical doctrine he had originated, and cultivated since 1852. In his autobiography, left unfinished at his death in 1899 and published in 1902 as *Sketches from the Life of Edward Frankland*, Frankland plays a substantial tribute to Kekulé, while at the same time firmly rebutting the eventual claims of the latter to prior discovery:*

The application of my theory of valency to carbon compounds, however, belongs substantially to Kekulé, whose brilliant application of this theory to carbon compounds generally and to benzol and its congeners especially, constitutes one of the most important epochs in the history of chemical science.

* *C.R. Acad. Sci.*, Paris, **58**, 510 (1864): see p. 56.

In preparing his collected works prior to 1877 for publication* with the laudable intention of rendering the earlier papers intelligible to students of that day, Frankland modernized where possible all formulae, and inserted consequential textual emendations. He thus largely obliterated the evidences, so interesting to the historian, of the slow-moving but ultimately drastic changes of chemical outlook occurring between 1852 and 1866, the period of his own most pregnant researches and one of the most critical periods in the history of chemistry (p. 34).

Hofmann, the first professor at the Royal College and Frankland's immediate predecessor, accepted the chair of chemistry at Berlin in 1864, and on the eve of his final departure for his native country issued *An Introduction to Modern Chemistry* (London, 1865). The work, dedicated appropriately to Sir James Clark, the Queen's physician and 'one of the principal founders of the Royal College', contributed much to the general knowledge and acceptance of the conceptions of valency, not least because it helped to initiate this convenient term. Frankland in all his publications consistently favoured 'atomicity', and so long as the plurality of atoms in elementary molecules remained generally unrecognized no confusion necessarily resulted. Kekulé, who came to fill Frankland's position as the doyen of valency theory, employed the more questionable term 'basicity' indiscriminately with 'atomicity'.

The following is a slightly abridged extract from pp. 168–9 of Hofmann's book:

We are in want of a good appellation to denote this atom-fixing power of the elements. The vague and rather barbarous expression, *atomicity*, has drifted into use for this purpose. This expression is faulty, because it is open to confusion with the atomic structure of elementary molecules. We shall escape this evil by substituting the expression *quantivalence* for *atomicity*, and designating the elements *univalent, bivalent, trivalent,* etc.

Not unnaturally the suggested term of Germanic proportions was soon shorn of its first two syllables. Hofmann also recommended small Roman numerals as superior suffixes, e.g. N^{III},

---

* *Experimental Researches in Pure, Applied and Physical Chemistry* (London, 1877).

C^{IV}, P^V, in place of the unsightly accumulation of dashes in current use for valency nomenclature. The new notation at once gained Frankland's approval, and continues as standard practice today.

On p. 183 of his book Hofmann touches explicitly on an aspect of valency only hinted at by Frankland. This is the absence of any general relation between valency and affinity.

The affinity of hydrogen and oxygen is evidently very great, but that of chlorine and oxygen is so feeble that its oxides are all liable to explosive decomposition, yet while two atoms of hydrogen combine with at most two atoms of oxygen, two atoms of chlorine can be united to seven atoms of oxygen.

The word *quantivalence* is employed to designate the particular numerical atom-compensating power in each of the elements, and this power must by no means be confounded with the specific *intensity* of their respective chemical activities. Thus nitrogen, phosphorus and arsenic are all trivalent, but this equality does not imply that they are all endowed with equal *avidity* for, say, oxygen or hydrogen. Because one *quadri*valent carbon atom can fix four atoms of hydrogen, of which one only is fixed by *uni*valent chlorine, we are not, therefore, to suppose that the attraction of carbon for hydrogen is necessarily more intense than that of chlorine, in the ratio 1/4. On the contrary the element which seizes the lesser proportion of hydrogen seizes it with by far the greater degree of energy.

2. *Archibald Scott Couper (1831-1892)*

Couper's genius flashed across mid-nineteenth-century chemistry during little more than one year and was then extinguished under cruel misfortune and total neglect of its achievements. In youth he had followed the Scottish tradition by devoting himself to the study of the classics and philosophy, first at Glasgow and then in Berlin, where, late in 1854, he became attracted to the natural sciences and particularly to chemistry. Following some training in practical analytical chemistry Couper moved in 1856 to Paris to work, like Kekulé, in Wurtz' laboratory, where his aptitude, although so lately realized, enabled him to engage in independent research, and to publish two papers in the field of benzene chemistry. It is ironical that in 1860, two years after its publication, both Kekulé and Kolbe failed to repeat Couper's work upon some reactions of salicylic acid: it was left to Anschütz in 1885 to prove the complete rectitude of Couper's statements. There can be no doubt that his third publication *On a New Chemical Theory* would in more fortunate circumstances have advanced him instantly to the front rank among contemporary chemists.

It would be strange indeed if in the ferment of new conceptions and hypotheses arising in chemistry during the mid-nineteenth-century years more than one practitioner did not independently give to the world similar, reformative ideas. It is a melancholy reflection that such consensus of opinion, which ought by its promise for the future to be a source of gratification to a philosophical mind, has led in science more often than in other pursuits to unedifying conflicts over the always ephemeral question of priority of publication. When Kekulé's now famous paper in the same field as Couper's appeared on 19 May 1858* Wurtz, as sponsor for its publication, had already kept Couper's MS in his hands for some months, and it was only through the hasty intervention of Dumas that Couper secured publication at the session of the Académie des Sciences on 14 June of that year.† Kekulé hastened to claim even so slight and dubious a priority in a Note to the Académie‡ read at the session of 30 August, and unctuously concluding: 'Loin de ma pensée... de soulever un débat de priorité, mais j'ai cru de mon devoir de constater l'originalité des vues exposées par moi.' Couper was certainly unwise to use in his paper so trenchant a style, lacking all that homage to established authority spread so lavishly (and so tediously) by Kekulé; and it could hardly be held against Wurtz that he had acted less than impartially by retaining so revolutionary a paper for further consideration before forwarding it for publication. But in the event Kekulé went forward eventually to claim for himself the prime discovery of the doctrine of valency (p. 56) while Couper's work and even his

---

* *Liebigs Ann.* **106**, 129 (1858).
† *C.R. Acad. Sci., Paris*, **46**, 1157 (1858).
‡ *Ibid.* **47**, 378 (1858).

name appeared irretrievably lost, until the timely resuscitation by Anschütz.*

Deeply distressed by such unfortunate events Couper took up a teaching post at Edinburgh, where he found a congenial colleague in his contemporary Alexander Crum Brown, later to succeed Playfair as Professor. In less than a year Couper suffered a breakdown of mental health from which he never again recovered sufficiently to defend his position or to participate in the further progress of chemical science.

The circumstances of the publication of the first French version of Couper's paper necessarily required some curtailment of detail, but within a short time he published full versions, in English† and in French.‡

In the early pages of his essay, which no doubt were those most alarming to Wurtz, Couper engages in a reasoned attack upon Gerhardt's theory of types, then replacing the older system of 'radicals'. In this critical exercise 'we discern clearly the lasting influence of his classical studies; his exposition is vivid, penetrating, and decisive, and his criticism does not lack the sting of sarcasm'.§ Gerhardt's theory‖ derived from the assumption that the constitution of organic compounds could be represented upon the basis of four essential 'types':

$$\left.\begin{array}{c}H\\H\end{array}\right\}, \quad \left.\begin{array}{c}H\\Cl\end{array}\right\}, \quad \left.\begin{array}{c}H\\H\end{array}\right\} O, \quad \left.\begin{array}{c}H\\H\\H\end{array}\right\} N.$$

For example, acetone was formulated as 'acetyl methide':

$$\left.\begin{array}{c}C_2H_3O\\CH_3\end{array}\right\},$$

which, as a *constitutional* formula, amply justifies Couper's severe criticism. Couper concludes that Gerhardt's theory, like its predecessors, 'is for the most part empirically true, that is to say

---

* Life and Chemical Works of Archibald Scott Couper, *Proc. roy. Soc. Edinb.* **29**, 194 (1908); *Ostwald's Klassiker*, no. 183 (1911).
† *Phil. Mag.* (4), **16**, 104 (1858).
‡ *Ann. Chim.* (*Phys.*), (3), **53**, 469 (1858).
§ Anschütz, *Ostwald's Klassiker*, **183**, 31 (1911).
‖ *Ann. Chim.* (*Phys.*), **37**, 33 (1853).

it is not contradicted by many of the facts of the science': but since it offers no *explanation* whatever of such facts it is not philosophically true.

The philosophical test demands that a theory be competent to explain the greatest number of facts in the simplest possible manner.... (Gerhardt) is led not to explain bodies according to their composition and inherent properties, but to think it necessary to restrict chemical science to the arrangement of bodies according to their decomposition, and to deny the possibility of our comprehending their molecular constitution.... Would it not be only rational, in accepting this veto, to renounce chemical research altogether?

There was at this time a situation in the physics of the properties of matter which might have afforded Couper a very apt analogy to that which he strove to inaugurate in chemistry. Only a year had passed since Clausius put forward the kinetic theory of the gaseous state, under which all the many previously discovered *empirical* laws of gases were explained.

Having shown so decisively (not to say derisively) the need for a new approach, Couper brings forward his theory as follows (text slightly abridged from the English version entitled *On a New Chemical Theory* and modern names inserted in parentheses where required):

It has been found that there is one inherent property common to all elements, denominated chemical affinity. It is discovered under two aspects:—
1. affinity of kind, 2. affinity of degree. The former is the special affinity of the elements, one for another, as carbon for oxygen, for chlorine, for hydrogen, etc. The second is the grades or limits of combination which elements display. For instance C^2O^2 and C^2O^4 are the degrees of affinity of carbon for oxygen; and the second degree is for carbon its ultimate affinity or combining limit. Affinity of degree may have only one grade, but generally has more than one.

Carbon has two highly distinguishing characteristics:
1. It combines with equal numbers of equivalents of hydrogen, chlorine, oxygen, sulphur, etc.
2. It enters into chemical union with itself. These two properties explain all that is characteristic of organic chemistry. The second property, being here signalized for the first time, requires supporting evidence. What is the link which binds together 4, 6, 8, 10, etc. equivalents of carbon and as many of hydrogen, or oxygen? You may remove perhaps all the hydrogen or oxygen and substitute so many equivalents of chlorine, etc. It is the carbon that is united to carbon. Also there is the conversion of the hydrocarbon C^4H^6

(ethane) into C^4Cl^6 and all the other countless instances of substitution of chlorine etc. which all tend in the same direction. This affinity (for itself), one of the strongest that carbon displays, is perhaps only inferior to its affinity for oxygen.

It is to be noted that Couper, while using Gmelin's equivalents, avoids the current term 'atomicity', which he tellingly replaces by 'affinity of degree': and that he clearly accepts variable valency as generally applicable, even to carbon.

On the tenth page of his paper, after presenting as examples of the application of the two postulates many formulae containing the current symbol C^2 ($C = 6$) for carbon, Couper writes 'There is no reaction found where it is known that C^2 is divided. It is only consequent to write, with Gerhardt, C^2 simply as C, it being then understood that 12 is the equivalent of carbon.' It therefore appeared excusable, as well as helpful to the modern reader, to substitute, in extracts from the earlier part, the usage Couper himself adopts in his later pages: in all subsequent extracts the symbol C takes its modern meaning ($C = 12$). For oxygen, however, while he recognized its 'combining limit' to be two, he everywhere retains as its symbol ...O...O... ($O = 8$), for reasons obscurely explained and apparently deriving from the dualistic theory.

A feature in the affinity of carbon is that it combines by degrees of two: thus C^2H^4 and C^2H^6; C^2H^4 and C^2H^5Cl. Hence it will always be impossible to isolate a compound of the form CH^3 or C^2H^5.... Carbon having only two grades of combination, of two atoms each, may legitimately furnish two grand types for all its compounds: the first nC M^4 and the second nC M^4–m M^2. As examples of the first type—

Methylic alcohol   Ethylic alcohol

In the latter instances when the combining limit (4) of each C is reduced by 3 in hydrogen, or in hydrogen and oxygen, there still remains a combining power of *one* to each C, which each expends in uniting with the other: the instances are therefore of the type nC M^4.

48

When two molecules of

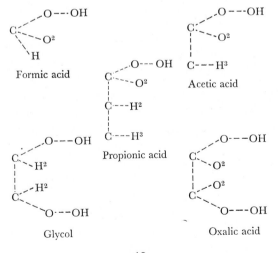

are set at liberty (by removing two molecules of water, HO, from two molecules of ethylic alcohol), the free affinities of the oxygen instantly produce their union, to form ethylic ether. Thus a property of oxygen causes the union of these two molecules, but the union of two molecules of CH^3 is due to a property of carbon.

At the close of his paper Couper declares his intention of dealing in a further communication with (unsaturated) bodies of the second type, but the opportunity was denied him. In all his graphical formulae in the English version of his paper Couper uses the broken line - - - to denote intramolecular links, but in the French version unbroken lines as well as the traditional bracket are admitted. In neither form, however, are his links made to signify explicitly the 'degree of combining power' involved, a further step to be taken later by Crum Brown, for a short time Couper's colleague at Edinburgh (p. 46).

The paper concludes by proposing formulae for all the main classes of saturated aliphatic compounds then known: for mixed ethers, normal propyl and butyl alcohols:

Formic acid

Acetic acid

Propionic acid

Glycol

Oxalic acid

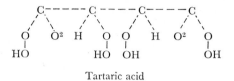

Tartaric acid

and finally and triumphantly the formula for (hydrated) glucose:

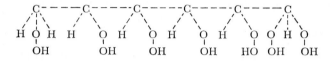

These formulae may all be transformed essentially into those in a modern text-book by the simple substitution everywhere of O (16) for O^2 or - - -O- - -O- - - (O = 8).

3. Friedrich August Kekulé (1829-1896)

Few chemists of his time can have been more travelled than Kekulé: of none could it more aptly be said 'He saw the cities of many men, and knew their minds'. From his early intention, on entering the University of Giessen in 1847, of becoming an architect of buildings he was soon diverted by the impress of Liebig's lectures to enter upon a course at length involving him in the architecture of molecules. Through the generosity of a relative he was enabled, in 1851, to move to Paris, where he came for a year under the influence of Gerhardt, and attended the lectures of Wurtz. In later years he recalled Liebig's cynical comment 'Gehen Sie nach Paris, da erweitern Sie Ihren Gesichtskreis, da lernen Sie eine neue Sprache, da lernen Sie das Leben einer Grossstadt kennen, aber Chemie lernen Sie dort nicht'. After two years spent in a private laboratory in Switzerland, he migrated in 1854 to the laboratory of St Bartholomew's Hospital in London, then under the direction of John Stenhouse: here by a curious coincidence Frankland was to succeed Stenhouse on returning to London in 1857, a year after Kekulé had returned to Germany. In spite of such a close convergence it has not been recorded that the two men ever met. During his two years' residence in London Kekulé was much in the company of Williamson and Odling, to whose influence upon him he pays tribute in his *Lehrbuch der organischen Chemie*. Kekulé returned to Germany in 1856 to spend two years as Privatdocent at Heidelberg under Bunsen, before being called to the chair of chemistry at Ghent in succession to Stas. It was soon after his arrival in Ghent that Kekulé entered upon what some have considered to be at once his most fertile and his most difficult undertaking—the organization of the Karlsruhe Conference of 1860. His wide international contacts ideally fitted him to be the impresario of the company of more than one hundred distinguished members, among whom Cannizzaro proved to be the principal actor.* From Ghent, in 1865, also issued his crowning achievement in theoretical chemistry—the constitution of benzene. By the time he returned again to his native country, in 1867 and in the 38th year of his age, to fill the professorship of chemistry at Bonn which he still held on his death in 1896, he had made known to the world all those brilliant conceptions which remain the permanent foundation of aromatic organic chemistry.

So far from attacking the current theory of types Kekulé set himself, in two papers,† to adorn it with emendations agreeable to the doctrine of valency. He adopted from the first and adhered to all his life a principle that the valency of every element 'is an atomic property as fixed and constant as its atomic weight'. Such a view is philosophically plausible and attractive, but does not bear the test of experiment, as Frankland had amply proved. It was fortunate that organic compounds are mainly composed of four light elements, carbon, nitrogen, oxygen and hydrogen, which fall into the small class of those possessing only one possible valency, but Kekulé encountered great difficulties later in the field of inorganic chemistry, where he propounded formulae often as unrealistic

* For a complete report of the Congress, with the names of those present, see R. Anschütz, *August Kekulé*, 1 (Berlin, 1929), 671.

† *Liebigs Ann.* **104**, 129 (1857); **106**, 129 (1858).

as those of Berzelius for organic bodies. Kekulé was always loth to acknowledge any competitors in originating or applying the doctrine of valency, and it is true that he stood practically alone in his obstinate support of a general principle of constant valency.

Throughout his communications on valency theory from 1857 onwards Kekulé used Gerhardt's atomic weights and Williamson's crossed symbols for those that differed from Gmelin's equivalents:

$$\text{H} = 1, \quad \Theta = 16, \quad \text{C} = 12, \quad \text{N} = 14.$$

Although he evidently preferred the term 'basicity', he used it indiscriminately with 'atomicity': to avoid confusion we use the former term exclusively in translated extracts of those papers which appeared in 1857 and 1858.

Ueber die gepaarten Verbindungen und die Theorie der mehratomige Radicale*

In a molecule of a compound the number of atoms of an element or of a radical bound to an atom of another element (or radical) is dependent upon the basicity or affinity numbers (Verwandschaftsgrösse) of the constituents of the molecule.

Elements fall in this respect into 3 principal groups:

1. Unibasic (I)   H, Cl, Br, K.
2. Dibasic (II) Θ, S
3. Tribasic (III)   N, P, As.

[in footnote]:

Carbon is easily shown to be tetrabasic: that is, 1 atom of carbon C (12) is equivalent to 4 atoms of H. The simplest compound of C with an element of Group I, for example, with H or Cl, is therefore CH_4 and CCl_4. Later I shall go further into this matter.†

There are therefore three principal types—

$$I+I, \quad II+2I, \quad III+3I,$$

that is, $\quad$ H H $\quad$ ΘH_2 $\quad$ NH_3.

More than one molecule of the same or different types may be united to give *multiple* or *mixed* types—

$$\begin{cases} \text{H Cl} \\ \text{H Cl} \end{cases} \qquad \begin{matrix} \text{H }\Theta \\ \begin{cases} \text{H} \\ \text{H }\Theta \end{cases} \\ \text{H} \end{matrix} \qquad \begin{cases} \text{H Cl} \\ \text{H }\Theta \end{cases} \\ \text{H}$$

* *Liebigs Ann.* **104**, 129 (1857). $\qquad$ † See p. 54.

A union within such multiple or mixed types can only occur when a polybasic radical, replacing 2 or 3 atoms of H, is present.

For example, *dibasic* radicals $\overset{''}{S}\Theta_2$, $\overset{''}{C_2H_4}$, $\overset{''}{C\Theta}$, $\overset{''}{C_2\Theta_2}$;

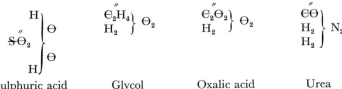

| Sulphuric acid | Glycol | Oxalic acid | Urea |

Triatomic radicals $\overset{'''}{P\Theta}$, $\overset{'''}{C_3H_5}$;

$$\left.\begin{array}{l}\overset{'''}{P\Theta}\\ H_3\end{array}\right\}\;\Theta_3 \qquad \left.\begin{array}{l}\overset{'''}{C_3H_5}\\ H_3\end{array}\right\}\;\Theta_3$$

Phosphoric acid   Glycerol

Gerhardt's formula for glycerol—

$$\left.\begin{array}{l}C_3H_5\Theta\\ H_2\end{array}\right\}\;\Theta_2$$

is inadmissible because the radical is only unibasic.

The terms monohydric, dihydric, and trihydric still used to classify alcohols derive from the number of molecules of water in the types appropriate to them. The formula quoted by Kekulé for sulphuric acid had been previously given by Williamson[*] and that for phosphoric acid by Odling,[†] both of whom recognized the role of the 'polyatomic' radical generalized by Kekulé. Glycerol had been correctly formulated by Wurtz.[‡]

Ueber die Constitution und die Metamorphosen der chemische Verbindungen und über die chemische Natur des Kohlenstoffs[§]

I consider it necessary, and now possible, to go back to the elements composing a compound in order to explain its properties. It is in my view no longer the most pressing task of the day to indicate groups of atoms which from their properties may be treated as radicals, and so to refer a compound to a few types, which have no other significance than as formal patterns. I believe rather that we must consider the constituents of the radicals them-

[*] *Quart. J. chem. Soc.* **4**, 350 (1852).   [†] *Ibid.* **7**, 1 (1855).
[‡] *Ann. Chim.* (*Phys.*), (3), **43**, 492 (1855).
[§] *Liebigs Ann.* **106**, 129 (1858).

selves, and the relations of radicals one to another; and from the nature of the elements infer both the nature of radicals and that of their compounds.

[As an example]

The radical $\overset{''}{S}\Theta_2$ is formed of three atoms, each dibasic: of the total of 6 'affinity units' 4 will be required to bind the three atoms together: the radical is therefore dibasic. The radical $\overset{''}{S}\Theta$ is also dibasic, since out of 4 'units' 2 are necessary to hold the atoms together. Hence for sulphuric and sulphurous acids—

$$\overset{''}{S}\Theta_2 + \left\{ \begin{matrix} H \\ H \\ H \\ H \end{matrix} \right. \begin{matrix} \Theta \\ \\ \Theta \end{matrix} \quad \begin{matrix} H \\ \\ H \end{matrix} = \overset{''}{S}\Theta_2 \left\{ \begin{matrix} \Theta \\ \Theta \end{matrix} \right.$$

$$\overset{''}{S}\Theta + \left\{ \begin{matrix} H \\ H \\ H \\ H \end{matrix} \right. \begin{matrix} \Theta \\ \\ \Theta \end{matrix} \quad \begin{matrix} H \\ \\ H \end{matrix} = \overset{''}{S}\Theta \left\{ \begin{matrix} \Theta \\ \Theta \end{matrix} \right.$$

All radicals can be analysed in a similar manner, even those containing carbon, but first a knowledge of the nature of carbon is required.

In the simplest compounds of carbon one atom of that element binds 4 atoms of a monobasic, and 2 atoms of a dibasic, element; the sum of the chemical units (*chemischen Einheiten*) of elements bound to one atom of carbon is 4: carbon is *tetrabasic*—

IV+4I	IV+(II+2I)	IV+2II	IV+(III+I)
ΘH_4	$\Theta\Theta Cl_2$	$C\Theta_2$	ΘNH
ΘCl_4		CS_2	
$\Theta H_3 Cl$			
ΘHCl_3			

It is doubtful whether this table in itself brought to his more critical contemporaries the conviction Kekulé designed. In a selection of 'the simplest compounds of carbon' not only is the simplest of all omitted, carbon monoxide, but also CH_4O, methyl alcohol, and CH_2O_2, formic acid, both of which provide *six* 'chemical units' to one atom of carbon. Only the first column of the table proved the quadrivalency of carbon without further assumptions, such as a preconceived principle of constant valency. For methane and its chlorine derivatives only one scheme of binding the elements is possible, but, for example, both $\overset{'''}{C}H$, N and $\overset{''}{C}$, NH were possible for hydrocyanic acid.

For substances containing several atoms of carbon it must be assumed that some at least of the atoms (other than carbon) are held by the affinity of the carbon, and that the carbon atoms themselves are held together so that a part of the affinity of one binds an equal part of the affinity of the other. The simplest and therefore the most probable case of the union of 2 carbon atoms is that one affinity-unit (*Verwandtschaftseinheit*) of one atom is bound to one of another. Of the total 8 units therefore 2 are employed in this way and the six remaining bind other elements: that is C_2 is *hexabasic*. [The maximum number of hydrogen atoms which can be bound by n carbon atoms is then shown to be $2n+2$.]

On comparing carbon compounds containing equal numbers of carbon atoms, and which are interconvertible by simple chemical changes, it appears that the carbon atoms retain their arrangement, and only the atoms attached to the carbon skeleton change.

In a very large number of organic compounds this 'simplest' arrangement of carbon atoms exists, but others contain so many carbon atoms in the molecule that in them a more compact (*dichtere*) arrangement of carbon atoms must be assumed. For example, benzene and its homologues, with all their derivatives, are characteristically distinguished from aliphatic compounds by a greater carbon content: in naphthalene, which contain an even higher proportion of carbon, a still more condensed form, with individual carbon atoms closer together, must be assumed.

[Footnote]
One is easily persuaded that the formulae of these (aromatic) compounds can be constructed from the 'next simplest' arrangement of the carbon atoms.

Series of hydrocarbons with more compact arrangement—

Ethylene	Propylene	Butylene	Amylene
C_2H_4	C_3H_6	C_4H_8	C_5H_{10}
	Benzene	Toluene	Xylene
	C_6H_6	C_7H_8	C_8H_{10}
		Naphthalene	
		$C_{10}H_8$	

After a summary of the atomicities of radicals, with examples, under the generalized headings:

uniatomic: $C_n H_{(2n+1)}$, $C_n H_{(2n-1)}O$,

diatomic: $C_n H_{2n}$, $C_n H_{(2n-2)}O$, $C_n H_{(2n-4)}O_2$,

triatomic: $C_n H_{(2n-1)}$,

the paper concludes in an unexpectedly philosophic vein:

I must remark that I myself place only a subordinate value upon the notions I have explained. Chemistry, being not yet an exact science, must be advanced by probable propositions, which prove useful in giving a simple generalized representation of the most recent discoveries, and are the means of reconciling future discoveries with old.

Sur l'atomicité des éléments*

In this communication Kekulé claims for himself alone the discovery of atomic valency—'C'est moi, si je ne me trompe, qui ai introduit en chimie la notion de l'atomicité des éléments' (cf. p. 42)—and states explicitly his own doctrine of its constancy for every element. To enable this doctrine to survive a steadily increasing body of experimental evidence against it, including that of Frankland, he puts forward a suitably flexible theory of 'molecular compounds', a term by which he appears prepared to describe all substances thermally dissociable.

The second paragraph of the paper carries a reminder of Berzelius.†

On sait que les corps élémentaires se combinent d'apres la loi des proportions constantes et la loi des proportions multiples. La première de ces lois trouve une explication parfaite dans la théorie atomique de Dalton; la seconde ne s'explique par la même théorie. Ce que la théorie de Dalton n'explique pas, c'est la question de savoir pourquoi les atomes des différents éléments se combine dans certains rapports plutôt que dans d'autres. J'ai cru expliquer l'ensemble de ces faits par ce que j'ai appelé l'atomicité des éléments.

Vouloir admettre que l'atomicité puisse varier...c'est confondre la notion de l'atomicité avec celle de l'équivalence: l'équivalent peut varier mais non l'atomicité. Les variations de l'équivalent doivent s'expliquer par l'atomicité.

Kekulé's successors of the present generation have been much exercised, and not unsuccessfully, to show that in the case of nitrogen its five oxides incorporate different manifestations of a constant tervalency; but in 1864 this could have been asserted only for organic compounds.

Les combinaisons dans lesquels tous les éléments sont tenus ensemble par les affinités des atomes qui se saturent mutuellement pourraient être nommées *combinaisons atomiques*. Ce sont les seules qui puissent exister a l'état de vapeur....Nous devons distinguer une seconde catégorie de combinaisons que je désignerai *combinaisons moléculaires*....L'attraction des atomes dans des molecules différents provoque le rapprochment et la juxtaposition des molécules, phénomène qui précède toujours les véritables decompositions chimiques. Or, il peut arriver que la réaction s'arrête à ce rapprochment;

* *C.R. Acad. Sci., Paris*, **58**, 510 (1864). † P. 26.

que les deux molécules se collent ensemble, formant ainsi un groupe doué d'une certaine stabilité....Parmi les combinaisons de ce genre je citerai—
Eléments triatomique: PCl_3, Cl_2; NH_3, HCl
Eléments biatomiques: $SeCl_2$, Cl_2; $TeBr_2$, Br_2
Eléments monoatomiques: ICl, Cl_2, etc.

Crystalline hydrates, and polyiodides are also quoted.

Kekulé seems to have been unaware that his conception of molecular compounds was a revival of an aspect of the dualistic theory discussed in detail by Berzelius some 40 years previously (see p. 23 ff). The chemistry of selenium and tellurium had been too little investigated in 1864 for him to know that the vapours of their tetrachlorides have normal vapour densities up to the temperature of red heat, but, on the contrary, it was certainly recognized that phosphorus oxychloride, $POCl_3$, could be distilled unchanged. However, in the succeeding years, until well into the present century, the term 'molecular compound' was to become a very convenient and ample cloak for concealing vacuities in chemical knowledge.

Kekulé published his earliest views on the constitution of benzene and its homologues, and the compounds derived from them, in two papers from Ghent: one to the Chemical Society of France* 'Sur la constitution des substances aromatiques' and another to the Belgian Royal Academy† 'Note sur quelques produits de substitution de la benzine'. A transcript, comprising both these papers, with some added experimental evidence relevant to his theory, appeared later, also from Ghent.‡

The ideas first introduced by these papers have been further adorned but not essentially changed under the impact of the most recent chemical theories: by universal consent they constitute the greatest and most fertile advance in the long history of organic chemistry. The inspired insight into the possibilities inherent in the power of carbon, acting with constant quadrivalency, to unite with itself not only laid the permanent foundation of 'aromatic' organic chemistry, but also led directly

* *Bull. Soc. chim. Fr.* (N.S.), **3**, 98–110 (Jan. 1865).
† *Bull. Acad. Belg. Cl. Sci.* (2), **19**, 551–63 (May 1865).
‡ *Liebigs Ann.* **137**, 129 (Feb. 1866).

to an understanding of the part played by the ethylenic linkage in general organic chemistry.

In introducing the first paper of 1865, *Sur la constitution des substances aromatiques*, Kekulé observes:

On n'a pas encore tenté, que je sache, d'appliquer les mêmes vues (de l'atomicité en général et de la tetratomicité du carbone en particulier) aux substances aromatiques. J'avais bien fait entrevoir, lorsque j'ai publié, il y a sept ans,* la théorie de la tetratomicité du carbone, que j'avais une idée toute formée à cet égard, mais je n'avais pas jugé à propos de la développer en détail.†

Perhaps not unaware that he was ushering in a new era in organic chemistry, Kekulé finally abandoned in this paper his long allegiance to the nomenclature of 'types', and, excepting the continued use of certain 'crossed' symbols (Θ, Θ) wrote single-line formulae such as are universal today. The following records, in translation, the essentials of the first paper:

(1) In attempting to form a conception of the atomic constitution of aromatic substances account must be taken of the following facts:

(*a*) The simplest aromatic compounds are always richer in carbon than aliphatic compounds (substances grasses).

(*b*) Homologues of benzene exist, differing in composition by $n\Theta H_2$.

(*c*) The simplest aromatic compounds contain at least 6 carbon atoms. Under energetic reagents more complicated aromatic substances give substances containing 6 carbon atoms.

(2) There is in all aromatic compounds a common group, a sort of nucleus, of 6 carbon atoms. Within this nucleus the carbon atoms are more condensed (plus condensés) than in aliphatic compounds. To this nucleus other carbon atoms become united in the same way as in aliphatic compounds.

(3) When many carbon atoms combine they may be united so that *one* of the 4 affinities of each atom saturates always *one* affinity of the neighbouring atom. It is also possible that several atoms of carbon may be united by 2 affinities between 2 atoms: further they may combine alternately by one and by 2 affinities, i.e.

$$1/1; \quad 1/1; \quad 1/1; \quad 1/1; \quad \text{etc.}$$

or $$1/1; \quad 2/2; \quad 1/1; \quad 2/2; \quad \text{etc.}$$

Six carbon atoms combined in the second way, considered as an open chain will have 8 affinities unsaturated; but if the two terminal atoms also combine, a *closed* chain, with 6 unsaturated affinities, results. [See formula, p. 60].

* *Liebigs Ann.* **106**, 129 (1858): see p. 53.   † See 'Footnote', quoted on p. 55.

(4) [As footnote in the original]

Ethylene can be considered as the first term of a series of closed chains. A hydro-carbon of the composition CH_2 (methylene) cannot exist: two affinities in the *same* atom cannot saturate each other [cf. Frankland's views, p. 37].

(5) Each of the six nuclear affinities can be saturated by a mono-atomic atom, or by one affinity of a polyatomic element; in the latter case the poly-atomic elements must bring in with them other atoms, to produce *side chains*.

(6) Examples of substitution.

(a) Mono-atomic elements: the theory indicates only one mono-chloro-benzene, but probably 3 isomers each of di-, tri-, and tetrachloro-benzene. An atom of chlorine substituted in benzene is exceptionally unreactive because it is surrounded (entouré) by carbon: in aliphatic chloro-bodies terminal chlorine is easily removed in double decompositions, but less easily when the chlorine atom is in the interior of the chain.

(b) Bi- and tri-atomic elements: Oxygen and nitrogen most commonly bring hydrogen in with them: the products may be supposed to contain the group ΘH or the group AzH_2.

$\mathrm{C_6H_5(\Theta H)}$	$\mathrm{C_6H_4(\Theta H)_2}$	$\mathrm{C_6H_3(\Theta H)_3}$
Phenic alcohol	Oxyphenic acid	Pyrogallic acid
$\mathrm{C_6H_5(AzH_2)}$	$\mathrm{C_6H_4(AzH_2)_2}$	$\mathrm{C_6H_3(AzH_2)_3}$
Aniline	Diamido-benzene	Triamido-benzene

The general properties of these bodies will differ from those of the alcohols and the aliphatic amines for the reasons given above.

This paragraph contains the first recognition by any organic chemist of the *hydroxyl* group—a much needed rationalization: in using the symbol Az for nitrogen Kekulé defers to a still sur-viving French preference.

(c) Nitro-bodies: The group $Az\Theta_2$ could combine with the nucleus either through oxygen or through nitrogen. It is probable that the oxygen atoms each combine with nitrogen by one of their affinities while the remaining affinities saturate each other.

In further sections the homologues of benzene are explained, and their isomerism: isomerism due to alternative substitution in a side-chain or the nucleus is exemplified by benzyl chloride and mono-chloro-toluene, and the oxidation of side-chains to the acid carboxyl group is discussed.

Kekulé's graphic formulae

At the end of the first paper in 1865, and of the later German version, Kekulé exhibits numerous examples of the type of graphic formulae he had used since about 1859, presumably in his lectures.* Elements were represented by elongated figures of which the lengths were made proportional to the valencies:

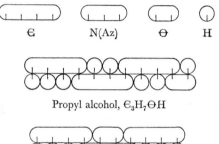

$$\text{Ꞓ} \qquad \text{N(Az)} \qquad \text{Ꝋ} \qquad \text{H}$$

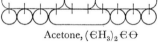

Propyl alcohol, $Ꞓ_3H_7ꝊH$

Acetone, $(ꞒH_3)_2 Ꞓ Ꝋ$

Benzene and its derivatives appear as containing a closed chain (or nucleus) of six carbon atoms, the closure being indicated by terminal arrows:

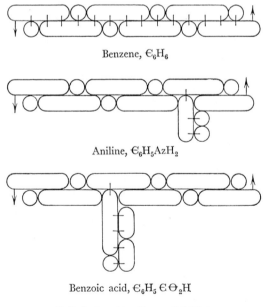

Benzene, $Ꞓ_6H_6$

Aniline, $Ꞓ_6H_5AzH_2$

Benzoic acid, $Ꞓ_6H_5 Ꞓ Ꝋ_2H$

* *Bull. Soc. chim. Fr.* **3**, 98 (1865).

Kekulé explains that his formulae were almost the same as he had seen Wurtz use in his lectures, and that he preferred them to those proposed by Crum Brown (pp. 39 ff.). Since his formulae are exasperating to draw, impossible to print except from plates, and capable of ambiguity,* Kekulé soon found himself alone in his preference. Reluctant to follow the lead of one who had shown his own to be untrustworthy, Kekulé deprived himself for some years of the help of any sort of graphic formulae in his papers.

In the atomic models he devised† Kekulé was more fortunate. After criticizing graphic formulae, necessarily wholly planar, for lack of spatial realism, he describes his carbon models 'instead of lying in one plane, the four units of affinity of the atom radiate from it in the directions of hexahedral axes, so that they terminate in the faces of a tetrahedron.... A model of this kind allows the union of 1, 2, or 3 units of affinity in a natural and obvious manner.' This was breaking ground untilled since Dalton's and Wollaston's time (p. 10), which less than a decade later and after further tending was to bear the fruit of a real molecular architecture.

Kekulé's second paper 'Note sur quelques produits de substitution de la benzine', published by the Belgian Royal Academy in 1865, is concerned with the question

whether the six hydrogen atoms in benzene all play the same rôle, or whether they have differing values. If the atoms are indistinguishable among themselves the cause of isomerism in benzene compounds must be sought in a difference of *relative* position between the atoms (or lateral chains) which replace hydrogen atoms in benzene: if not, then isomerism can also arise from the *absolute* position of substituents, and the number of possible isomeric forms will be greatly increased.

It is not necessary here to follow in detail Kekulé's theoretical discussion (which lacks his usual clarity) of a problem which he ultimately solved by experiment.

The vast expanse of the field of research opened by Kekulé's theory of benzene began at once to attract to his laboratory, first

---

* Crum Brown, *Proc. roy. Soc. Edinb.* **5**, 429 (1864).
† *Z. Chem.* **3**, 217 (1867).

at Ghent and after 1867 at Bonn, collaborators, nearly all of whom rose in after years to become leaders of the science. After six years of well-organized attack upon benzene problems Kekulé could report* that no example of isomerism among benzene derivatives had come to light which could not be completely explained 'by the difference in relative positions of the atoms substituted for hydrogen'. This conclusion might be concisely expressed by claiming that benzene possessed hexagonal symmetry. Kekulé realized clearly that another problem replaced that solved, for his formula of benzene admitted only of trigonal symmetry. Many chemists had already attempted solutions of this conflict between theory and experiment, but the most arresting came again from Kekulé's genius.*

The atoms in molecules must be assumed to be in continual motion, but nobody has, to my knowledge, expressed opinions about the form of this motion. Chemical principles demand that atoms in a molecule retain their relative positions (in derselben relativen Anordnung verharren) and therefore always return to a median position of equilibrium (stets zu einer mittleren Gleichgewichtslage zurückkehren). Valency may be interpreted as the *relative* number of collisions in a unit of time that an atom suffers from other atoms. In a molecule composed of 2 univalent atoms linked to a bivalent atom the latter suffers in a unit of time twice as many collisions as are inflicted upon each of the univalent atoms.

In the case of benzene (I) the collisions between atom 1 and its immediate neighbours in a unit of time may be expressed in the order 2, 6, h, 2. In the succeeding time interval the order may equally well be 6, 2, h, 6, which is as probable as the first. But at the end of the second interval the formula of benzene has changed to that in II.

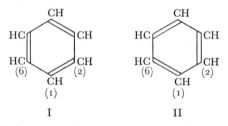

| I | II |

In an indefinitely long succession of very short time intervals we can understand that the carbon atom (1) will come to be similarly related to each of its neighbours (zu seinen beiden Nachbarn genau in derselben Beziehung steht), and therefore that the positions 1,2 and 1,6 are equivalent.

* *Liebigs Ann.* **162**, 77 (1872).

VALENCY AS AN EXPLICIT DOCTRINE

Anyone acquainted with the modern theory of benzene, which originated about the year 1930, might be tempted to see in these speculations of 60 years before a genuine example of clairvoyance. In March 1890, the 25th anniversary year of the publication of his benzene theory, the German Chemical Society, under the presidency of A. W. Hofmann, honoured Kekulé in Berlin with celebrations, lasting over several days, of a magnificence that has remained unique in the annals of the Society.* The young Emperor Wilhelm II, who had been a student at Bonn and is said to have attended Kekulé's lectures, hastened to take the 'Kekulé-Feier' under his express patronage. Although then physically very frail, Kekulé replied to the laudations showered upon him with a long and characteristically lively speech, during which he disclosed how often his structural ideas had first come to him in dreams, and in particular how in this way he had first hit upon his formula for benzene:

Während meines Aufenthaltes in Gent in Belgien bewohnte ich elegante Junggesellenzimmer in der Hauptstrasse. Mein Arbeitszimmer aber lag nach einer engen Seitengasse und hatte während des Tages kein Licht. Für den Chemiker, der die Tagesstunden im Laboratorium verbringt, war dies kein Nachteil. Da sass ich und schrieb an meinem Lehrbuch: aber es ging nicht recht; mein Geist war bei anderen Dingen. Ich dreht den Stuhl nach dem Kamin und versank in Halbschlaf. Wieder gaukelten die Atome vor meinen Augen. Kleinere Gruppen hielten sich diesmal bescheiden im Hintergrund. Mein geistiges Auge, durch wiederholte Gesichte ähnlicher Art geschärft, unterschied jetzt grössere Gebilde von mannigfacher Gestaltung. Lange Reihen, vielfach dichter zusammengefügt: Alles in Bewegung, schlangenartig sich windend und drehend. Und siehe, was war das? Eine der Schlangen erfasste den eigenen Schwanz und höhnisch wirbelte das Gebilde vor meinen Augen. Wie durch einen Blitzstrahl erwachte ich; auch diesmal verbrachte ich den Rest der Nacht um die Consequenzen des Hypothese auszuarbeiten.

Lernen wir träumen, meine Herren, dan finden wir vielleicht die Wahrheit... aber hüten wir uns, unsere Träume zu veröffentlichen, ehe sie durch den wachenden Verstand geprüft worden sind.

(*During my time in Ghent I occupied smart bachelor quarters in the main street. My study however lay against a side lane and was deprived of light during the day, but this*

---

* Bericht über die Feier der Deutschen Chemischen Gesellschaft zu Ehren August Kekulé's, *Ber. dtsch. chem. Ges.* **23**, 1265 (1890).

was little disadvantage to a chemist whose daylight hours were spent in the laboratory. There I sat writing my text-book, but to little avail: my thoughts were upon other things. I turned my chair to the fire and dozed. Again the atoms gambolled before my eyes, but this time the small groups kept demurely in the background. My mind's eye, sharpened by many previous experiences, distinguished larger structures of diverse forms: long series, closely joined together: all in motion, turning and twisting like serpents. But see, what was that? One serpent had seized its own tail, and this image whirled defiantly before my eyes. As by a lightning flash I awoke: and again spent the rest of the night working out the consequences of this idea.

Let us learn to dream, gentlemen, and then we may find the truth . . . but let us beware of making our dreams public before they have been approved by the waking mind.)

Kekulé was probably inclined to give too much weight to the approval by the 'awakened mind' of a hypothesis first apprehended in fantasy. The notion that every element possesses a unique and fixed valency is one that must appeal very strongly to a philosophical mind, yet it does not survive the test of experiment. Kekulé's lifelong adherence to such an hypothesis, in later years quite alone, is paralleled only by Dalton's equally persistent denial of Gay-Lussac's law of gaseous combination. The genius of each was somewhat foiled by a too strongly deductive cast of mind: it may also be significant that neither was an outstanding experimenter.

After 1876 Kekulé suffered a rapid deterioration in his physical health, and although his mental powers remained scarcely affected, a decline in the tempo of work in the Bonn laboratory and consequently of the numbers of young chemists seeking to collaborate with the master was inevitable. A decade later a new focus of work on the problem of benzene constitution had arisen in the laboratory at Munich, where Adolf von Baeyer had succeeded Liebig in 1875. Unlike Kekulé, von Baeyer had little inclination towards working at the head of a school of younger men: his own genius lay in experimentation rather than in the imagination of new theories, and almost the whole corpus of investigations on the benzene problem issuing from Munich between 1888 and 1893 originated, like the brilliant work on indigo preceding it, from his private laboratory. During those years benzene theory probably gained at least as much from Baeyer's refutation, in crucial tests aimed with unerring markmanship, of theories rivalling that of Kekulé, as it did from his more positive achievements.

For convenience in prescribing the positions of substituents in benzene the carbon atoms are numbered consecutively 1–6 in the hexagonal formula. Investigations inspired by Kekulé had proved conclusively that only three types of di-substituted benzene derivatives are possible, termed *ortho-*, *meta-* and *para-* compounds, and represented in Kekulé's formulation as follows:

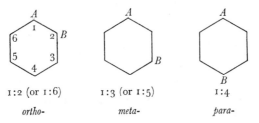

1:2 (or 1:6)	1:3 (or 1:5)	1:4
ortho-	*meta-*	*para-*

W. Körner, who after working with Kekulé at Ghent became Professor of Organic Chemistry in the Scuola di Agricultura at Milan, had also shown how a di-substituted compound $C_6H_4X_2$ may be referred conclusively to its class, provided only that the six hydrogen atoms in benzene are assumed to be equivalent.* On introducing a third substituent isomeric tri-substitution derivatives must be formed according to the following scheme:

Di-substituted compound	Tri-substituted products
Para-	One compound only
Ortho-	Two isomeric compounds
Meta-	Three isomeric compounds

The two formulae surviving as possible alternatives to that of Kekulé were both designed to avoid the three double bonds which, even in the 'oscillating' form he suggested, many chemists found unacceptable in a hydrocarbon as stable as benzene.

* 'Studi sull'isomeria delle cosi dette sostanze aromatiche a sei atomi de carbonio', *Gazzetta*, **4**, 305 (1874).

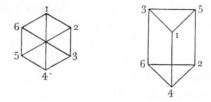

The first, by replacing one member of each of Kekulé's three double *ortho*-linkages by single *para*-linkages, automatically confers full hexagonal symmetry, but at the cost of raising the novel and severe problem of interpreting the 'crossing' of C—C links. In the second proposal the six CH groups occupy the corners of a regular triangular prism and only trigonal symmetry is achieved, but no special difficulty arises in understanding the C—C links. Each formula was first suggested by Ad. Claus* but he himself preferred and supported the 'diagonal' formula, while A. Ladenburg became the principal advocate of the prism type.† It may readily be demonstrated, by applying Körner's method to the prism formula, that the consecutive numbering of carbon atoms appropriate to the hexagonal formulae of Kekulé and Claus must be transferred in the way shown in the diagram. Hence equivalent *meta*- and *ortho*-relationships, each six in number, lie along the sides of the parallel triangular faces, and along the diagonals of the vertical faces respectively, while the vertical edges provide the three equivalent para-relationships.

Baeyer opened up a new field of 'hydroaromatic' chemistry in an exhaustive study of the products of reduction by hydrogen of the benzene nucleus.‡ Benzene itself is resistant to reducing agencies available before 1900 (when Sabatier and Senderens introduced their catalytic method) but the three di-carboxylic acids derived from it, terephthalic (I), isophthalic (II) and phthalic acid (III), are readily attacked by sodium amalgam and water.

* *Theoretische Betrachtungen und deren Anwendung zur Systematik der organische Chemie* (Freiburg, 1867).

† *Ber. dtsch. chem. Ges.* **2**, 141, 272 (1869).

‡ *Liebigs Ann.* **245**, 103 (1888); **251**, 257 (1889); **256**, 1 (1890); **258**, 1, 145 (1890); **266**, 169 (1892); **269**, 145 (1892); **278**, 88 (1893).

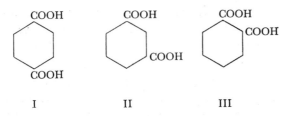

I   II   III

A typical course of action, discovered and used by Baeyer, is exemplified in the following scheme, resulting in the reduction of phthalic acid to its hexahydro derivative:

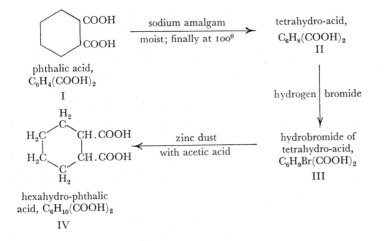

By similar means hexahydro-terephthalic and hexahydro-isophthalic acids were prepared.

Baeyer found that hexahydro-phthalic acid was converted into its anhydride,

on merely melting it,* as is the parent phthalic acid: while hexahydro-isophthalic acid gave an anhydride only very slowly

* The hexahydro-acid exists in two stereoisomeric forms (*cis-* and *trans-*), which Baeyer separated: both yield the anhydride.

even when heated with strong dehydrating agents, and hexa-hydro-terephthalic acid remained stable even under such conditions. There appeared therefore to be no doubt that after reduction of phthalic acid the original *ortho*-relationship of the carboxyl groups survived, as is required by the above formulation based upon Kekulé's formula for phthalic acid. When the prism formula is adopted for that acid, three of its nine bonds between the nuclear carbon atoms have to be removed so as to leave the six carbon atoms united by single bonds in a six-membered ring. This process can be carried out in only two ways:

Bonds removed	Orientation of carboxyl groups in reduced acid
1:4, 2:6, 3:5 (or 1:3, 2:5, 4:6)	*Meta-*
1:5. 2:4, 3:6	*Para-*

There exists no way of preserving the *ortho*-orientation.

The results of his exploration in hydroaromatic chemistry were again the basis of Baeyer's refutation of the 'diagonal' formula. On bromination hexahydro-terephthalic acid (I) yields a mono-bromo- and a di-bromo-derivative (II and III):

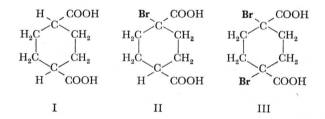

I   II   III

The derivative (II) was converted by alcoholic potassium hydroxide into the tetrahydro-acid (IV), which, treated alternatively with hydrogen bromide or bromine gave the products (V) or (VI) on p. 69. (V) and (VI) proved to be isomeric, and not identical, with (II) and (III) respectively: the constitution of the dibromo-derivatives (VI) and (III) was in this way confirmed.

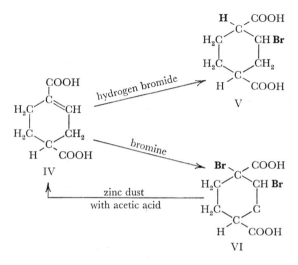

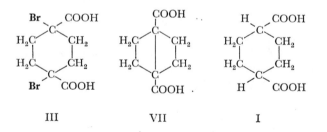

When treated with zinc dust and acetic acid these two di-bromo-compounds behaved quite differently: (VI) regenerated the tetrahydro-acid (IV), while the isomer (III) gave, not a tetrahydro-acid (VII) to be expected from the diagonal formula, but the original hexahydro-acid (I).

After his five years' study of their di-carboxy derivatives Baeyer turned his attention to the parent hydrocarbons, dihydro-, tetrahydro- and hexahydro-benzene, now more usually named cyclohexadiene, cyclohexene, and cyclohexane respectively, which he succeeded in preparing in quantity for the first time. The starting-point was diketohexamethylene (II, p. 70) obtained by hydrolysis from synthetic succinylo-succinic ester (I).

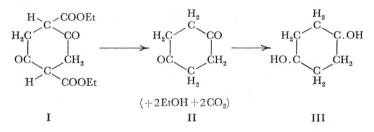

$$(+2\,\text{EtOH}+2\,\text{CO}_2)$$

I   II   III

The diketo-compound is easily reduced to 1:4-dihydroxy-cyclohexane (III), from which the three reduction products of benzene may be obtained by simple and long-familiar methods.*

Stohmann and Langbein in Leipzig determined the heats of combustion of specimens of the hydrocarbons supplied by Baeyer,† who quotes their results as follows:*

	Heat of combustion (kcal.)	Differences
Benzene	779·8	
		68·2
Dihydrobenzene	848·0	
		44·0
Tetrahydrobenzene	892·0	
		41·2
Hexahydrobenzene	933·2	
		58·0
(Hexane)	991·2	

and adds the following comment, drawn largely verbatim from Stohmann and Langbein's paper:

Benzene cannot contain three equivalent double linkages, at least of the kind which occur in the aliphatic series. The most stable linkages occur in benzene, before reduction, and the least stable in dihydro- and tetrahydro-benzene. In hexahydrobenzene a larger stability is regained, which however does not reach that of the original benzene.

The order of the increases in heat of combustion in the third column of the table brings out in quantitative form what had already been evident in all Baeyer's researches: namely, that the 'aromatic' properties peculiar to benzene are not gradually lost during its progressive reduction, but suddenly disappear in the first stage of reduction to dihydrobenzene. Conversely it may be asserted that it is the entry of the third ethylenic linkage into the hexagonal formula that originates these properties.

* *Liebigs Ann.* **278**, 88 (1893).   † *J. prakt. Chem.* **48**, 447 (1893).

As the well-chosen 'Festredner' at the Kekulé-Feier in 1890 (p. 63) Baeyer thus opened his address: 'In his first expositions of 1865–6 Kekulé set up two formulae for the benzene ring, represented as

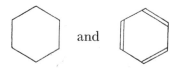

but gave his preference to the second which has become generally known as the Kekulé formula....We shall see that *both* formulae are justified; that is, that the two taken together explain the known facts better than any other formula.' Later in his speech he summed up his own work upon terephthalic acid: 'By the study of its reduction products it has been proved that the benzene ring in terephthalic acid behaves as a ring containing three double bonds.'*

In concluding the 'Festrede' Baeyer disclosed his own views upon the constitution of benzene and its derivatives.

The explanation of the discontinuity of properties at the formation of the third double linkage has now become the crux of the benzene theory...the benzene ring approaches a limiting state (*Grenzzustand*) which I may call 'ideal' benzene, in which the six CH groups are bound together with exceptional strength (*ausserordentlicher Festigkeit*), so that the ring seems to become completely symmetrical, and the fourth valency of the carbon atoms disappears from our cognizance (*für unsere Wahrnehmung verschwindet*). In this state the atoms appear to be tervalent, and we may express it by single lines joining the six CH groups. In order to signify the special meaning of these lines, which differ from those we ordinarily use to symbolize single linkages, I propose for 'ideal' benzene the formula I have termed 'centric'—

The two limiting states of benzene are consequently represented by the Kekulé formula and the centric formula....The nature of the ring in any particular derivative is that of some intermediate state, and for ordinary usage therefore the Kekulé formula well serves our purpose.

* See further, p. 73.

As Baeyer acknowledged at another time, H. E. Armstrong had previously put forward the 'centric' representation, expressly stating that it could formulate only benzene itself and probably not any derivative.* It is difficult to perceive that the centric formula was more than a recognition of the current insolubility of the final benzene problem: and it was curious that a representation so completely immune from the test of experiment should be sponsored by a supreme experimenter.

In 1893 Johannes Thiele, then at Halle, accepted Baeyer's invitation to join him at Munich in the then vacant office of 'Extraordinarius für organische Chemie'. Baeyer's immediate object was to gain a colleague able and willing to relieve him largely of the management of a large laboratory† but it is difficult to believe that he had no intuition of the part Thiele was to play towards solving the 'final' benzene problem.

In a long communication published in 1899‡ Thiele first drew attention to unexpected features in the reactivity of carbon compounds containing pairs of double linkages separated by only one single bond, which he termed a 'conjugated system'. Thus muconic acid (I) is readily reduced (by sodium amalgam and water at room temperature) to dihydro-muconic acid (II)

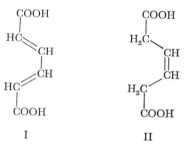

I   II

and Baeyer had earlier emphasized the close analogy between this change and the reduction of terephthalic acid by the same reagents (III and IV)

* *J. chem. Soc.* **51**, 264, footnote (1887).
† J. Thiele (1865–1918), obituary, *Ber. dtsch. chem. Ges.* **60**, 75 (1927).
‡ *Liebigs Ann.* **306**, 125 (1899).

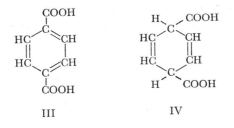

III   IV

in support of his view that the ring in terephthalic acid comprised three double linkages (see quotation, p. 71).

Thiele cited many other examples of such behaviour of 'conjugated systems', amongst which a number involved a carbon to oxygen or carbon to nitrogen double linkage, and so established a new general principle: that the atoms terminating a conjugated system are specially reactive towards additive agencies, and particularly to reduction. J. Thomsen had in 1887* calculated from thermochemical data that the heat of formation of a double binding between carbon atoms is considerably less than the heat of formation of two single links: Thiele saw in this fact a physical basis for his theory of 'partial valencies'.

He postulated that in forming a double linkage neither atom fully saturates its valencies, each still preserving a 'partial' active valency.

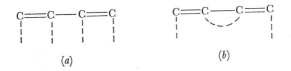

(a)   (b)

Such partial valencies are indicated in (a) and (b) by broken lines: the central pair mutually saturate each other, leaving the two terminal atoms in a state of enhanced activity towards additive reagents. Thiele regarded benzene as a unique example of *continuous* conjugation containing three double

* *Z. phys. Chem.* **1**, 369 (1887).

linkages but nevertheless possessing exceptional stability owing to the complete mutual saturation of all the partial valencies.

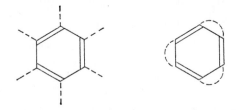

Baeyer had already shown that the dihydrobenzene (I) and its derivatives show active unsaturation towards reducing agents, while the isomer (II) and its derivatives do not.

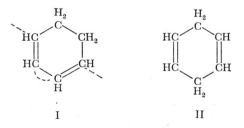

As long as benzene remained the unique example of 'continuous' conjugation a direct test of Thiele's theory remained impossible, but in 1911–1913 Willstätter synthesized the hydrocarbon *cyclo-octatetrene*

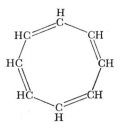

by methods leaving no doubt about its constitution.* The new hydrocarbon, however, proved to be extremely reactive and dissimilar to benzene in all respects. Benzene therefore retained

* *Ber. dtsch. chem. Ges.* **44**, 3423; **46**, 517.

its isolated position, the ultimate cause of which could not have been foreseen until the development of the most recent theories and experimental methods. These together have at last shown that the uniqueness of benzene arises from a stereochemical cause—all its twelve atoms lie in the same plane. Kekulé's original inspiration that benzene is a regular hexagon has thus been more exactly confirmed than he could have expected.

BIOGRAPHICAL AND OTHER REFERENCES

Sketches from the Life of Edward Frankland, 1825–1899. Edited and concluded by M.N.W. and S.J.C. (London, 1902.)

Experimental Researches in Pure, Applied and Physical Chemistry. E. Frankland. (London, 1877.)

August Kekulé. R. Anschütz. Vols. I and II. (Berlin, 1929.)*

'Life and Chemical Works of Archibald Scott Couper'. *Proc. R.S. Edinb.*, **29**, 194 (1908). (See also, *Ostwald's Klassiker*, no. 183, 1911.)

An Introduction to Modern Chemistry. A. W. Hofmann. (London, 1865.)

Chemical Society Memorial Lectures:
A. W. Hofmann, *J. chem. Soc.* **69**, 575 (1896).
F. A. Kekulé, *J. chem. Soc.* **73**, 97 (1898).
S. Cannizzaro, *J. chem. Soc.* **101**, 1677 (1912).
A. Ladenburg, *J. chem. Soc.* **103**, 1871 (1913).
A. v. Baeyer, *J. chem. Soc.* **123**, 1520 (1923).

Obituaries:
W. Körner, *J. chem. Soc.* **127**, 2975 (1925).
J. Thiele, *Ber. dtsch. chem. Ges.* **60**, 75 (1927).

* Vol. II contains complete reprints of all Kekulé's scientific papers, addresses and articles.

4

NEWLANDS, MENDELEEF
AND L. MEYER

The incorporatian of valency in a chemical system

Before the periodic law was formulated atomic weights were purely empirical numbers; so that the magnitude of the equivalent and the atomicity (or the value in substitution possessed by an atom) could only be tested by critically examining the method of determination, but never directly by considering the numerical values themselves: in short we were compelled to move in the dark, to submit to the facts, instead of being masters of them.

MENDELEEF, Faraday Lecture of the Chemical Society, 1889

The law of the periodicity of the elements, based jointly upon the atomic properties of weight and valency, was, after the atomic theory, the greatest contribution of nineteenth-century chemistry to natural science. It can be imagined graphically by setting out the steadily increasing atomic weights along an abscissa, with corresponding valencies on the ordinate, when the latter appear as quantities fluctuating in a regular (but not uniform) pattern. The law provided a means of classification which unified and rationalized the whole effort of chemical investigation, and was the necessary forerunner of our present knowledge of atomic constitution, and consequently of the cause of valency and of chemical union. Conversely the success of the periodic system of classification rapidly enforced the universal acceptance of true atomic weights, upon which alone the system could rest: and not less important the recognition of a principle of valency applicable to all problems of molecular constitution.

Classification of the elements discovered by the methods of

his science is rightly a preoccupation of the chemist, and from the time of Lavoisier's broad and qualitative separation into metals and non-metals various more detailed systems of classification had been advocated. Since Dalton's atomic theory bequeathed only weight as a numerical property of the isolated atom such attempts were necessarily confined to seeking arithmetical relations between atomic weights, which themselves had no fixed and rational values until the reforms after 1860. In the following decade, not only was valency established as a second numerical atomic property, but a system of assured atomic weights based on the principles of Avogadro and Cannizzaro was rapidly replacing the former arbitrary values. Numerical data for the minimum number of atomic properties required for a basis now becoming available, it was almost inevitable that more than one chemist should perceive the means of a rational classification. Although Mendeleef in 1869* set up the periodic system more completely and convincingly than others he freely admitted† the independent claims of de Chancourtois (Professor of Geology at the Ecole des Mines, Paris)‡ and Newlands§ to the prior discovery of a law of periodicity. Lothar Meyer‖ used as his criterion of periodicity 'atomic volume', defined as atomic weight/specific gravity. Such an apparently simple and uncontroversial means for its demonstration strongly influenced the acceptance of the idea of periodicity.

1. John Alexander Reina Newlands (1837-1898)

A Londoner by birth, Newlands was a student at the Royal College of Chemistry under the professorship of Hofmann, who doubtless gave encouragement to his speculative turn of mind.

* *J. Soc. phys. chem. russ.* **1**, 60 (1869); *Liebigs Ann.* (Suppl.), **8**, 133 (1872).
† *J. chem. Soc.* **55**, 638 (1889).
‡ *C.R. Acad. Sci.*, Paris, **54**, 757, 840, 967; **55**, 600; **56**, 263, 479; **63**, 24 (1862–1866).
§ *Chemical News*, **12**, 83 (1865).
‖ *Liebigs Ann.* (Suppl.), **7**, 354 (1870).

After holding several teaching appointments he became the chief chemist to a sugar refinery at Victoria Docks, London.

Largely because they were content to rely, as evidence of periodicity, upon the recurrence, with increasing atomic weight, of similar but *qualitative* chemical properties, neither de Chancourtois nor Newlands gained the recognition due to them. Newlands, however, proposed a definite 'law of octaves'. 'The (ordinal) numbers of analogous elements generally differ either by 7 or some multiple of seven: in the nitrogen group between N and P are 7 elements: between P and As, 14: between As and Sb, 14: and lastly between Sb and Bi, 14.'* He also prepared a table† in which the quotient of atomic weight by ordinal number (now termed *atomic number*) is seen to be nearly constant round the value 2·5 for most of the then known elements: a result prophetically foreshadowing modern developments.

One of Newlands's tables (*Chemical News*, 1865) is shown below:

No.	No.	No.	No.	No.	No.	No.	No
H 1	F 8	Cl 15	Co, Ni 22	Br 29	Pd 36	I 42	Pt, Ir 50
Li 2	Na 9	K 16	Cu 23	Rb 30	Ag 37	Cs 44	Tl 53
G¹ 3	Mg 10	Ca 17	Zn 24	Sr 31	Cd 38	Ba, V 45	Pb 54
B 4	Al 11	Cr 18	Y 25	Ce, La 32	U 40	Ta 46	Th 56
C 5	Si 12	Ti 19	In 26	Zr 33	Sn 39	W 47	Hg 52
N 6	P 13	Mn 20	As 27	Di,² Mo 34	Sb 41	Nb 48	Bi 55
O 7	S 14	Fe 21	Se 28	Rh, Ru 35	Te 43	Au 49	Os 51

¹ Glucinum = beryllium.   ² Didymium = Praseodymium + Neodymium.

In March 1866, Newlands read a paper on 'The Law of Octaves, and the Causes of Numerical Relations among Atomic Weights' to the Chemical Society: it was derided by the members present and no order was given for its publication. The senior Chemical Society in Europe thus scornfully threw away the high honour of first publishing the periodic law, which was, ironically but also perhaps significantly, to be given to the world from the pages of the first volume of the journal of the youngest Society. Newlands took advantage of an alternative

* *Chemical News*, **12**, 83 (1865).   † *Ibid.* **12**, 94 (1865).

but less far-reaching publication in the *Chemical News*, edited from its foundation in 1859 until 1906 by W. Crookes (later Sir William Crookes). In a volume dated 1884* under the title 'On the Discovery of the Periodic Law' he collected and reprinted all his papers so published, and by this means soon brought for himself a long-delayed recognition from the Royal Society. In presenting to him the Davy Medal at the Anniversary Meeting of 1887, the President, Sir George Stokes, recalled that five years earlier the Medal had been awarded jointly to Mendeleef and Lothar Meyer, and continued

While recognizing the merits of the chemists of other nations we are not to forget our own countrymen: and accordingly the Davy Medal for the present year has been awarded to Mr J. A. R. Newlands for his discovery of the Periodic Law of the Chemical Elements. Though in the somewhat less complete form in which the law was enunciated by him, it did not at the time attract the attention of chemists, still in so far as the work of the two foreign chemists was anticipated, the priority belongs to Mr Newlands.

2. *Dmitry Ivanovich Mendeleef (1834-1907)*

Mendeleef was a native of Tobolsk and the youngest of a family of fourteen. Through the untiring efforts of his widowed mother he was enabled to receive instruction in natural science at the Central Pedagogic Institute in St Petersburg. At the very early age of 22 he became Privatdocent in the University of that city. After sojourns in Paris under Regnault, and in Heidelberg, he returned to St Petersburg to occupy successively the chair of chemistry at the Technological Institute, and, in 1866, the chair of general chemistry at the University. Three years later, in the first volume of the *Journal of the Russian Physical and Chemical Society* he published the paper 'On the Relation of the Properties of the Elements to their Atomic Weights', which ushered in the periodic system of classification. In 1890 he resigned his professorship for political reasons, but held the Government post of Director of Weights and Measures until his death in 1907. He

* London, E. and F. N. Spon.

paid a number of visits to England and was well known to many English chemists. In 1889 the Chemical Society conferred upon him its highest honour—the Faraday Medal, which entailed the delivery of the Faraday Lecture, for which he chose the title 'The Periodic Law of the Chemical Elements'.*

Mendeleef enunciated and developed the periodic law in three communications to the Russian Physical and Chemical Society: in March and November 1869, and in December 1870.† The last paper, which contains the most detailed account of the law and its consequences, was reprinted in verbatim translation in *Liebigs Ann.* (Suppl.) **8**, 133 (1871), and all the papers are printed as German translations in *Ostwalds Klassiker*, no. 68. In the last paper Mendeleef boldly asserted his confidence in his 'natural system' by venturing to predict not only the existence but a full range of properties of three then unknown elements. As their names indicate, the hope expressed by Berzelius in introducing his chemical symbols, that 'chemistry would recognize no frontiers', was fulfilled in the discovery of these elements: gallium, by Lecoq de Boisbaudran, 1875: scandium, by Nilson, 1879: germanium, by Winkler, 1886. 'When, in 1870, I described the properties which such elements ought to possess, I never hoped that I should live to mention their discovery to the Chemical Society of Great Britain as a confirmation of the exactitude and the generality of the periodic law.'‡ As is well known the predicted properties were confirmed in detail. Newlands five years earlier indicated vacant places for the same elements, but with less confidence, and without risking the prediction of properties.

The following freely rendered extracts, with commentary, are drawn mainly from the Russian paper of 1871.

Cannizzaro's prescription for fixing atomic weights demanded volatile compounds of the element concerned: it was therefore least easily applied in the case of metals, and in this field older

---

* *J. Chem. Soc.* **55**, 634 (1889).

† *J. phys. chem. Soc. russ.* **60** (1869); *ibid.* **14** (1870); *ibid.* **21** (1871); Mendeleef records the dates of his publications in a footnote to a paper *Ber. dtsch. chem. Ges.* **4**, 348 (1871).

‡ Faraday Lecture.

methods continued in use. These derived principally from (a) Dulong and Petit's law of atomic heats, (b) the law of isomorphism, and (c) the chemical character of the oxides, which were classified as follows:

$MO(M_2O)$, strongly basic;

M_2O_3, weakly basic;

MO_2, M_2O_5, MO_3, acidic oxides of strength increasing with oxygen content.

The example of beryllium (treated by both Mendeleef and Newlands) provided probably the most severe test of the law of periodicity:

This divergence of opinion (between Be (II) = 9 and Be (III) = 13·5) lasted for years, and I often heard that the question as to the atomic weight of beryllium threatened to disturb the generality of the periodic law, or at any rate to require some important modifications of it.... It is most remarkable that the victory of the law was won by the very observers who had previously discovered a number of facts supporting the trivalency of beryllium.*

The specific heat of beryllium is 0·425, giving, according to the atomic heat law, an atomic weight of approximately 14: the oxide is only weakly basic, and its crystal form (hexagonal system) is quite different from that of magnesium oxide (cubic system), but sufficiently resembles that of alumina (corundum) to suggest a close relationship: and beryllium chloride like aluminium chloride is volatile. The evidence for association with aluminium rather than with magnesium as demanded by the periodic law appeared very strong. The final settlement in favour of the periodic system came from the adoption of Cannizzaro's principle: the vapour density of the volatile chloride allowed only the atomic weight 9·0 and therefore bivalency of the element.

Although thoria and zirconia act as strong bases, the system offers places for these metals only as quadrivalent elements, with oxides ThO_2, and ZrO_2. On the other hand uranium oxide is found to be weakly basic and was accordingly given the formula U_2O_3: the system demands however a position in which this element is sexavalent, and its oxide UO_3. Only the equivalent

* Faraday Lecture, 1889.

81

of indium (38) was known in 1869, and Mendeleef uses this case to demonstrate the power of his system in deciding its atomic weight and valency. It could occupy only the gap in the series of the table between Cd (II) = 112 and Sn (IV) = 118, and was therefore a tervalent element: Mendeleef himself redetermined its specific heat (0·055) to confirm this position.

The oxides of the newly discovered metals in cerite, cerium, 'didymium', and lanthanum, all possessed strongly basic properties but the chemistry of the metals differed widely from the chemistry of those with oxides MO already placed in the system. The uncertainty felt by Mendeleef in absorbing these elements into his classification endured almost to the present day, and the problem was finally solved only through a knowledge of atomic structure. On the other hand yttrium, the oxide of which was supposed to be YO, filled the gap in the series between Sr^{II} and Zr^{IV}, both of which possessed strongly basic oxides; the equally basic oxide of yttrium has therefore the form Y_2O_3, and the element is tervalent.

The concluding section (6) of Mendeleef's paper constitutes an historically most valuable critical survey of valency theory as it was held about 1870. The principal issues raised in this section are summarized as follows:

(1) Kekulé's division of chemical compounds into 'atomic' and 'molecular' types is artificial, arbitrary and unsound: even ammonium chloride and phosphorus pentachloride are relegated to the status of 'molecular compounds'. No practical test exists by which the two categories may be sharply separated: the formation of 'molecular' compounds from molecules already completely stable; their decomposition on vaporization; the minor changes in reaction capacity ensuing from their formation, are all unavailing, since from them $PtCl_4$, $KClO_4$, etc. will be styled 'molecular compounds'.

When, however, the conception of molecular compounds is not accepted valencies cannot necessarily be established from hydrogen or chlorine compounds alone: $PtCl_4$ and SiF_4 cannot represent the valency limits for platinum and silicon, for these halides combine respectively with HCl and HF to form PtH_2Cl_6

Mendeleef's second table (1871)

Series	Group I — R_2O	Group II — RO	Group III — R_2O_3	Group IV RH_4 RO_2	Group V RH_3 R_2O_5	Group VI RH_2 RO_3	Group VII RH R_2O_7	Group VIII — RO_4
1	H = 1							
2	Li = 7	Be = 9·4	B = 11	C = 12	N = 14	O = 16	F = 19	—
3	Na = 23	Mg = 24	Al = 27·3	Si = 28	P = 31	S = 32	Cl = 35·5	—
4	K = 39	Ca = 40	— = 44	Ti = 48	V = 51	Cr = 52	Mn = 55	Fe = 56, Co = 59 Ni = 59, Cu = 63
5	(Cu = 63)	Zn = 65	— = 68	— = 72	As = 75	Se = 78	Br = 80	—
6	Rb = 85	Sr = 87	?Yt = 88	Zr = 90	Nb = 94	Mo = 96	— = 100	Ru = 104, Rh = 104 Pd = 106, Ag = 108
7	(Ag = 108)	Cd = 112	In = 113	Sn = 118	Sb = 122	Te = 125?	I = 127	—
8	Cs = 133	Ba = 137	?Di = 138	?Ce = 140	—	—	—	—
9	(—)	—	—	—	—	—	—	—
10	—	—	?Er = 178	?La = 180	Ta = 182	W = 184	—	Os = 195, Ir = 197 Pt = 198, Au = 199
11	(Au = 199)	Hg = 200	Tl = 204	Pb = 207	Bi = 208	U = 240	—	—
12	—	—	—	Th = 231	—		—	—

83

and SiH_2F_6: many very stable 'double cyanides' may also be cited.

(2) Only 17 out of the 63 elements at present discovered are known to combine with hydrogen and only *one* hydride is given for each: but with other elements, such as chlorine or oxygen, many compounds may be formed: carbon yields CH_4 but not CH_2 while with oxygen both CO and CO_2 are formed: other examples are $SnCl_2$, $SnCl_4$: $HgCl$ and $HgCl_2$: PCl_3 and PCl_5. Can valencies therefore be assigned from hydrogen compounds alone? If from water and hydrogen chloride we take oxygen and chlorine to have maximum valencies of II and I respectively, we are compelled to accept such formulae as

$$K—O—O—O—O—Cl, \quad K—O—O—O—Cl, \quad K—O—Cl$$

and must explain why the longest oxygen chain is the most stable. Platinum oxide which is accepted as $O{=}Pt{=}O$ is much less stable than $KClO_3$.

(3) Can valency be regarded as a fixed and unalterable atomic property? Taking sulphur and oxygen to exhibit a constant bivalency the formulae for SO_2 and SO_3 must be

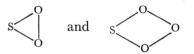

but sulphur dioxide is chemically analogous to carbon dioxide, with formula $O{=}C{=}O$. The compound $S(Et)_3I$ has been discovered, and the analogous elements tellurium and molybdenum form $TeCl_4$ and $MoCl_4$: sulphur can be *quadrivalent*. Again the analogous element tungsten gives WCl_6 and sulphur may well be *sexavalent* in sulphur trioxide. Likewise chlorine is heptavalent in Cl_2O_7. We may ask what guarantee exists that hydrogen and oxygen have no higher valencies?

(4) Convinced that the current 'laws' of valency need revision and re-statement, Mendeleef proposes three principles 'firmly based upon experimental knowledge of chemical compounds'.

(a) *The principle of substitution.* When a molecule is decomposed into two parts, the one part is equivalent to the other (and each can replace the other); e.g.

$$H---Cl, \; HO---H, \; C---H_4, \; CH_2---H_2, \; CH_3---NH_2$$

(b) *The principle of limits.* One at least of the molecules formed from the decomposition of a larger molecule can combine with amounts of other elements equivalent to the second molecule produced in the decomposition; e.g. C_2H_4 produced from C_2H_6O by elimination of H_2O combines with HCl or Cl_2 which are each equivalent to H—OH.

(c) *The periodic principle.* The highest compounds of an element with hydrogen, oxygen or other equivalent elements are determined by the atomic weight of the element, of which they form a periodic function.

(5) Examples of the application of these principles.

Hydrides provide only *four* forms:

$$RH, \; RH_2, \; RH_3, \; RH_4;$$

one hydrogen atom is replaceable by an equivalent residue from its own or another form: CH_3—CH_3, CH_3—NH_2, etc. The order of sequence of such substitution is immaterial: NH_2CH_3 is identical with CH_3NH_2. Further substitution can lead to two different products: CH_2ClNH_2 and CH_3NHCl for example. Thus homology and isomerism are explained.

With *oxygen* the following forms are found:

$$R_2O, \; RO, \; R_2O_3, \; RO_2, \; R_2O_5, \; RO_3, \; R_2O_7, \; RO_4.$$

For no element does the sum of the hydrogen and oxygen equivalences exceed *eight*:

Oxide	RO_4	R_2O_7	RO_3	R_2O_5	RO_2
Hydride	None	RH	RH_2	RH_3	RH_4 (no hydrides RH_5 or higher type are known)

Acids are produced by substituting equivalent amounts of H and OH in known oxy-forms:

$$\text{from } SO_3: SO_2(OH)_2, \; SO_2H(OH), \; SO_2H_2.$$

Chlorine forms correspond to the oxygen forms:

$$RO_2 - RCl_4; \ R_2O_3 - RCl_3; \ RO - RCl_2;$$

but forms often occur below the limiting form, and may then resemble the hydrogen form:

$TeCl_4/TeO_2$ but not $TeCl_6/TeO_3$,

ICl_5/I_2O_5 but not ICl_7/I_2O_7 (as in periodates),

$AsCl_3/As_2O_3$ but not $AsCl_5/As_2O_5$.

No chlorine forms higher than an oxygen form are known.

Complete molecules may combine:

$$SiF_4 \qquad SiF_4.2HF,$$

and many similar compounds R_2XCl_6 are formed from elements with oxide forms RO_2. $PtCl_4$ combines with $2RCl$, $2NH_3$, $4NH_3$, etc. Many so-called double salts, such as RBF_4 and R_3AlF_6 are very stable. The 'double' compounds

$$\left. \begin{array}{c} BCl_3 \\ POCl_3 \end{array} \right\}, \quad \left. \begin{array}{c} AlCl_3 \\ POCl_3 \end{array} \right\}, \quad \left. \begin{array}{c} AlCl_3 \\ AlCl_3 \end{array} \right\},$$

may be regarded as of similar constitution.

To characterize an element two data are necessary, which must be found through observation, experiment and comparisons with other elements, viz. the true atomic weight, and the true valency. Since the periodic law clearly expresses the mutual dependence of these data it affords the possibility of determining the valency when the atomic weight is given.

Not the least of Mendeleef's services to chemistry was his publication of the first text-book of chemistry to be based throughout on the periodic system of classification. In the preface to the fifth Russian edition he wrote

This book was written during the years 1868–1870...I wished to show in an elementary treatise on chemistry the palpable advantages gained by the application of the periodic law, which I first saw in its entirety in the year 1869 when I was engaged in writing the first edition of the book.*

* *The Principles of Chemistry*, English translation by G. Kamensky and A. J. Greenaway (London, 1891).

3. *Julius Lothar Meyer (1830-1895)*

Lothar Meyer's father and maternal grandfather were both physicians, and his mother had regularly assisted her father in his practice: Meyer could therefore hardly escape an early training for the medical profession, which appears to have encouraged an innate scientific versatility; his interests ranged from the subject of his *Doktorsarbeit* for the M.D. degree at Würzburg—'On the Gases of the Blood'—to the independent enunciation of the Periodic Law, and in both of these fields he was a pioneer. His quantitative researches on the relation of the gases oxygen, carbon dioxide and monoxide to the haemoglobin of the blood initiated the modern biochemistry of this subject. As Privatdocent in chemistry at Breslau he wrote *Die modernen Theorien der Chemie*, published in 1864, which had a powerful influence in establishing the views of Cannizzaro (p. 28). By a curious coincidence, in 1868 Lothar Meyer succeeded as Professor at the Karlsruhe Polytechnicum, Weltzein, who had been Kekulé's principal coadjutor in organizing the Karlsruhe Congress, and while the Polytechnicum served as a military hospital during the Franco-Prussian War he reverted as Army Surgeon to his family profession. In 1876 he succeeded Fittig at Tübingen, where he died in 1895.

Lothar Meyer's choice of 'atomic volume' as a means of displaying periodicity harks back to Dalton's attempt to compare the 'sizes' of the ultimate particles of common gases, from which his first table of atomic weights (1803) emerged as an offshoot.* For solid and liquid elements 'atomic volume' is a very complex quantity, depending upon temperature, physical state, mode of aggregation (crystal structure) as well as on absolute atomic volume. However, it is now known that chemically similar elements usually adopt similar crystal structures, and it is to this fact that Lothar Meyer mainly owed his success.

Although by his atomic volume curve Meyer probably succeeded more immediately than Mendeleef in arousing the

* Palmer, *Valency, Classical and Modern* (Cambridge, 1944), p. 2.

general interest of chemists in the idea of the periodicity of the elements, the demonstration lacked much of the finality of Mendeleef's exposition; in particular it provided, as Mendeleef pointed out, no decisive means of assessing the completeness of the natural atomic order. Consequently Meyer failed to predict missing elements, nor would he readily countenance any emendation of existing 'atomic weights'.

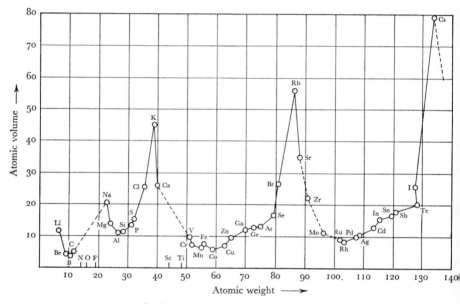

Lothar Meyer's atomic volume curve.

BIOGRAPHICAL AND OTHER REFERENCES

On the Discovery of the Periodic Law. J. A. R. Newlands. (London, 1884.)
Chemical Society Memorial Lectures:
D. I. Mendeleef, *J. chem. Soc.* **95**, 2077 (1909).
J. Lothar Meyer, *J. chem. Soc.* **69**, 1403 (1896).
Faraday Lecture, Chemical Society: D. I. Mendeleef, *J. chem. Soc.* **55**, 634 (1889).

5

PASTEUR, LE BEL AND VAN'T HOFF

Valency as a directional property

I am inclined to think, that when our views are sufficiently extended, to enable us to reason with precision concerning the proportions of elementary atoms, we shall find the arithmetical relation alone will not be sufficient to explain their mutual action, and that we shall be obliged to acquire a geometrical conception of their relative arrangement in all three dimensions of solid extension.

If there be three particles (combined with one) they might be arranged with regularity, at the angles of an equilateral triangle in a great circle surrounding the single spherule.... When the number of one set of particles exceeds in the proportion of four to one, then a stable equilibrium may take place if the four particles are situated at the angles of the equilateral triangles composing a regular tetrahedron.

WILLIAM HYDE WOLLASTON, *Phil. Trans.* **98**, 96 (1808)

Es muss unbedingt die Möglichkeit zugegeben werden dass bei Körpern von *gleichen* Strukturformel, welche letztere nichts anderes als die Verbind-ungs*folge* der das Molekül constituirenden Atome darstellen soll, doch noch Verschiedenheit in gewissen Eigenschaften als Ergebniss von verschiedene *räumlich* Anordnung der atoms innerhalb des Moleküls auftreten konnen.

JOHANNES WISLICENUS, *Liebigs Ann.* **165**, 47 (1873)

(*The possibility must certainly be granted that differences in certain properties can occur even between bodies of* identical *structural formulae, which represent only the* order *of combination of the constituent atoms of the molecule. Such differences must be the result of different* spatial *arrangement of the atoms within the molecule.*)

The speculations of Wollaston and the fact that Dalton preceded Kekulé in the use of ball and pin molecular models by about half a century might suggest that in their time some general notions of a three-dimensional chemistry were commonly entertained. Unfortunately (but perhaps inevitably) the creed of the organic chemists of the first half of the nineteenth century that a chemical formula could not and indeed should not represent more than an epitome of reactions condemned as heretical any more imaginative conceptions. Even Pasteur, after the publication in 1860 of lectures to the French Chemical Society on the interpretation of the most entrancing experiments of the century, was forced to admit: 'Few memoirs have had a better reception than mine, and yet I have had many proofs that they are little regarded.' Pasteur's exposition of the optical activity of dissolved organic compounds was so exhaustive and so compelling that it is now almost incredible that another fourteen years elapsed until the further exertions of Wislicenus, van't Hoff, and Le Bel at length awakened the chemical world to its significance.

1. Louis Pasteur (1822-1895)

Like Kekulé, Pasteur was the son of a distinguished soldier, Chevalier de la Légion d'Honneur and decorated on the field of battle by Napoleon I. Born at Dôle, he amply justified the early promise recognized by his teachers at the neighbouring College of Besançon by his career at the École Normale in Paris. Here he attended the lectures of Balard, the discoverer of bromine, and of Dumas at the Sorbonne: his acquaintance with a contemporary, Delafosse, who had been assistant to Haüy, led him to seek instruction also in crystallography. He began his work upon tartaric acid in 1844 immediately after graduating, when he became assistant to Balard. In 1860, upon the invitation of the Chemical Society of Paris, he delivered two lectures 'On Molecular Asymmetry' summarizing his results, published by the Society in a separate volume under the title *Leçons de Chimie professées en 1860* (Paris, 1861). A translation by Alexander Scott was published in the Alembic Club Reprints in 1897.

Pasteur began his first lecture by recalling the antecedents of his own work on the tartaric and paratartaric (racemic) acids and their salts. From 1813 onwards Biot had devoted himself to the study of what is now termed *optical activity*, that is the power shown by certain transparent media of rotating the plane of polarization of polarized light transmitted through them. At first his work was confined to crystal media, such as Iceland spar (calcite) or quartz: he soon made the discovery that some specimens of the latter rotated to the right (clockwise) and others equally to the left. In 1815 Biot laid the foundation of all the wide-ranging study since undertaken by chemists in the field of optical activity, by discovering that certain organic compounds, such as cane-sugar, camphor, turpentine, or tartaric acid rotate the plane of polarization even in *the liquid* or *dissolved* state, and he realized that the activity is not merely a property of a mode of arrangement in crystals, as had been presumed heretofore, but must be ascribed in the organic compounds to the ultimate particles themselves.

All the plane faces characteristic of a crystal are in general related by symmetry conditions: a given type of face occurs as one member of a set of equivalent faces, such as for example the six principal faces of a cubic crystal. However, in some crystals minor faces or *facets* are observed, each of which lacks its symmetrical counterpart, and such facets are termed *hemihedral.** Crystals exposing such hemihedry may occur in two non-superposable forms related as an object to its mirror-image, or as right-hand to left-hand.† Häuy, many years previously, noted such dual forms among specimens of quartz, and Biot confirmed the suggestion of Herschel in 1820 that these might correspond with the two sorts of crystals he had separated on the ground of opposed rotatory activity.

In 1841 M. de la Provostaye had published a beautiful piece of work on the crystalline forms of tartaric and paratartaric acids and their salts. I made a study of this memoir. I crystallized tartaric acid and its salts, and as the work proceeded I noticed that a very interesting fact had escaped the learned physicist. All the tartrates I examined gave undoubted evidence of hemihedral facets....These results, which may be summed up 'the tartrates are hemihedral and that in the same sense', has been the foundation of all my later work....Being very anxious to find some experimental support for

---

* This term has since become disused by most crystallographers.

† Now termed *enantiomorphous* forms.

the still speculative view of a possible inter-relation between the hemihedry and the rotatory power of the tartrates my first thoughts were to see whether other crystallizable and active organic compounds had hemihedral crystal forms. I met with the success I expected. On the other hand paratartrates were inactive to polarized light and none of them exhibited hemihedry. Thus the idea of the inter-relation of hemihedry and rotatory power gained ground.

During the year 1844, when Pasteur had almost completed his curriculum at the École Normale and was soon to begin work on the tartrates, Biot communicated to the Académie des Sciences on behalf of Mitscherlich the following Note:

The sodium ammonium salts of the (dibasic) paratartaric and tartaric acids have identical chemical composition, crystalline forms, specific weight, and double refraction. But when dissolved in water tartaric acid rotates the plane of polarization, while paratartaric acid is indifferent, yet here the nature and the number of atoms, their arrangement and distances are the same in the two substances compared.

Upon this Pasteur commented 'I thought at once that Mitscherlich was mistaken on one point. He had not observed that his double tartrate was hemihedral while his paratartrate was not. If this is so, his result is no longer extraordinary, and I should have the best test of my idea of the inter-relation of hemihedry and the rotatory phenomenon....

I hastened therefore to re-investigate the crystalline form of Mitcherlich's two salts. I found that the tartrate was hemihedral, but strange to say the *paratartrate was hemihedral* also. Only, the hemihedral faces which in the tartrate were all turned the same way, were, in the paratartrate inclined sometimes to the right and sometimes to the left.... I carefully separated the right-handed from the left-handed crystals and examined their solutions separately in the polarimeter. I then saw with no less surprise than pleasure that the crystals hemihedral to the right deviated the plane of polarization to the right, and those of the opposite sense deviated the plane to the left. When I took an equal weight of each of the two kinds of crystals the mixed solution was indifferent towards polarized light.'

Thus Pasteur describes the first 'resolution', as the process is now called, of a racemic mixture into its enantiomerides, and the discovery of *l*-tartaric acid, which does not occur as such in nature. He recalls how Biot, at first sceptical of his results, but fully realizing their great importance if true, insisted upon a repetition of the experiments in his presence at the Collège de France, and at the successful conclusion exclaimed 'Mon cher enfant, j'ai tant aimé les sciences dans ma vie que cela me fait battre le cœur'. Biot's emotion was understandable for he must have seen in his youthful disciple's work the possible culmination of his vain striving during 30 years to impress chemists with the importance of molecular asymmetry to the progress of their science.

During his lecture, Pasteur captivated his audience by exhibiting the experiment of mixing concentrated solutions of equal weights of the two tartaric acids, when the para-acid crystallizes spontaneously and is found to be in all respects identical with the natural product.

We know, on the one hand, that the molecular structures of the two tartaric acids are asymmetric, and on the other, that they are rigorously the same, with the sole difference of showing asymmetry in opposite senses. Are the atoms of the right acid grouped on the spirals of a dextrogyrate helix, or placed at the summits of an irregular tetrahedron? A resumé of the general relations between properties and corresponding atomic arrangements is as follows:

(1) When the atoms of organic compounds are grouped asymmetrically the crystalline form of the substance manifests this molecular asymmetry in non-superposable hemihedry.

(2) The existence of the asymmetry betrays itself also by the optical rotative property.

(3) When the non-superposable molecular asymmetry is realized in opposite senses the chemical properties of these identical but inverse substances are rigorously the same. It follows that this mode of contrast does not affect the ordinary play of chemical affinities.

Many years later van't Hoff gave the explanation* why Mitscherlich's double salt behaved on crystallization differently from the racemates (paratartrates), the crystals of which Pasteur had found to be devoid of hemihedral facets. Below 27° C. the solubility of the racemate becomes greater than the equal solubilities of the d- and l-salts, which are therefore the stable forms below this temperature: above 27° the converse is the case. For other racemates it happens that the racemate is stable at a *lower* temperature, being then *less* soluble than the d- or l-form.

* *Z. phys. Chem.* **1**, 173 (1887); *Ber. dtsch. chem. Ges.* **31**, 2206 (1898).

2. Jacobus Henricus van't Hoff (1852-1911) and Joseph-Achille Le Bel (1847-1930)

van't Hoff left on record that at his school in Rotterdam the lower forms were taught that HO was the correct chemical formula for water, but in the higher part of the school word went round that another formula, H_2O, was gaining ground. Perhaps thus so deterred from studying such an unsettled science as chemistry appeared to be, he devoted his attention mainly to mathematics and natural history until, after a year at the University of Leyden, his natural bent asserted itself. He repaired to Bonn, where Kekulé had recently arrived as Professor: the latter, perhaps recalling his own youth, soon advised a further move, to Wurtz' laboratory at the École de Médecine in Paris. Here van't Hoff gained the friendship of J.-A. Le Bel. A nephew of Boussingault, who, like Liebig, exerted himself to promote a knowledge of 'soil science', Le Bel, being a man of ample means, had no need to seek academic posts, and continued for many years to work in the laboratory in the nominal position of professor's assistant, first to Balard, and later to Wurtz.

In June 1874 van't Hoff left Paris to return to Holland, whence in the September following, at the age of 22, he issued his famous pamphlet 'An attempt to extend into space the present structural formulae of chemistry' ('Voorstel tot Uitbreiding den Structuurformules in de Ruimte'). In November of that year Le Bel's paper on the same subject appeared.* On the remarkable coincidence van't Hoff wrote twenty years later

That shortly before this (the publication of the two papers) we had been working together in Wurtz' laboratory was purely fortuitous: we had never exchanged a word about the tetrahedron there, though perhaps both of us had cherished the idea in secret. To me it had occurred the year before, after reading Wislicenus's paper on the lactic acids....On the whole Le Bel's paper and mine are in accord; historically the difference lies in this, that Le Bel's starting point was the researches of Pasteur, mine those of Kekulé.†

* *Bull. Soc. chim.* **22**, 337 (1874).   † Ref. bibl. (6) (see p. 106).

It is probable that during his time at Bonn van't Hoff's attention was drawn to Kekulé's models (p. 61).

Like Pasteur's lectures fourteen years before, and perhaps partly because it was written in van't Hoff's native language, his pamphlet received at first little if any recognition, which it gained ultimately in full measure, from two causes: the ardent support of Wislicenus, who, after procuring a German translation, himself wrote a preface for its publication in 1877: and secondly the scurrilous jeers of the aging Kolbe, which, being everywhere read for amusement, further extended the needed publicity. In the volume *Dix Années dans l'Histoire d'une Théorie* (1889)* after the reprint of his original essay of 1874, van't Hoff writes: 'C'était le début. Il n'y a que dix années ecoulées depuis. M. Kolbe est mort dejà et, comme par un jeu fatal du sort, c'est M. Wislicenus qui lui a remplacé a l'Université de Leipsig.'

Before being called in 1876 to the chair of chemistry, mineralogy and geology at the University of Amsterdam, van't Hoff spent two years in a teaching post at the Veterinary College at Utrecht, whence issued the monograph 'Ansichten über die organische Chemie', containing a suggestion of a model for the stereochemistry of nitrogen. For his Inaugural Address at Amsterdam he characteristically chose the topic 'The Rôle of Imagination in Science'. To the sorrow of his colleagues he accepted in his declining years an invitation to Berlin to apply the laws of physical chemistry he had himself discovered to the problems of the Stassfurth salt industry. In 1901 he received the first Nobel Prize for Chemistry.

The following is a synopsis of the paper by Le Bel:

Sur les rélations qui existent entre les formules atomiques des corps organiques et le pouvoir rotatoire de leurs dissolutions

Although Pasteur and others established completely the relation between molecular dissymmetry and rotatory power, and showed that if dissymmetry exists only in the structure of the crystal then only the crystal will be optically active, we still lack definite rules to predict whether a solution of a compound will possess rotatory power. We only know that usually (but not always) the

* Ref. bibl. (5) (see p. 106).

derivatives of an active compound are themselves active. Reasoning upon purely geometrical assumptions I have been able to formulate a much more general rule.

(1) *First general principle*

If in a compound MA_4, where M is a simple or complex radical, and A a monatomic atom, three atoms of A are replaced by three radicals (simple or compound) all differing among themselves and different from M, then the substance so obtained will be dissymmetrical and optically active.

Exceptions to this principle:

(*a*) If MA_4 possesses a plane of symmetry containing the four atoms A, then substitution will not yield a dissymmetrical product.

(*b*) The last radical substituted for A may be composed of the same atoms as the residue of the molecule into which it enters.

('*Le dernier radical substitué à A peut être composé des mêmes atomes que tout le reste du groupement dans lequel il entre.*')

Then the effects of these two like groups of atoms upon polarized light may either compensate each other, or be superimposed: in the former case the substance will be inactive.

(2) *Second general principle*

If in MA_4 only two radicals R^1 and R^2 are substituted, the result may be symmetrical or dissymetrical according to the constitution of MA_4.

(*a*) if MA_4 has a plane of symmetry containing the two atoms of A substituted, then no dissymmetry can arise.

(*b*) if on substituting R^1, R^2 or R^3 for one, two or three atoms of A in MA_4 only one compound results in each case, then in MA_4 the atoms of A must occupy the four apices of a regular tetrahedron, and no bi-substituted product of MA_4 can be active.

Applications to saturated compounds of the aliphatic series

All saturated aliphatic bodies can be derived from methane, CH_4, by substitution. Since methane gives only one product by bi- or tri-substitution it must have a regular tetrahedral configuration. Therefore if in the (graphic) formula of a compound a carbon atom can be found that is united to four different groups the compound is active; but if a carbon atom is united to two identical radicals the compound is inactive.

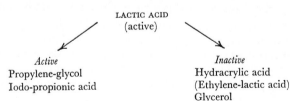

LACTIC ACID
(active)

Active	*Inactive*
Propylene-glycol	Hydracrylic acid
Iodo-propionic acid	(Ethylene-lactic acid)
	Glycerol

[And similarly for malic acid.]

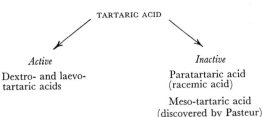

Mesotartaric acid is a case of the compensation mentioned under 'First principle, Exception (b)'.

Application to unsaturated (ethylenic) aliphatic compounds
(*'Corps gras à deux atomicités libres'*)

(1) If in ethylene, C_2H_4, the hydrogen atoms, are all coplanar then no tri-substituted derivative can be active.

(2) Whether the hydrogen atoms are coplanar or not no di-substitution in ethylene with *like* radicals (as in fumaric and maleic acids) can produce activity. If the hydrogen atoms in ethylene are not co-planar two isomers, one symmetrical and the other dissymmetrical, will be produced by two *different* substituents. A study of the optical properties of the amylenes $CH_2 = C (CH_3) (C_2H_5)$ and the isomeric $(CH_3) CH = CH (C_2H_5)$ should decide whether the hydrogen atoms in ethylene are or are not co-planar.

Theorem

When a dissymmetrical substance is formed in a medium containing only symmetrical substances, equal quantities of the two isomers of inverse symmetry will be produced.

The following is a synopsis of the essay by van't Hoff:

Essai d'un système de formules atomiques à trois dimensions, et la relation entre le pouvoir rotatoire et la constitution chimique qui en découle

1. (*a*) If the four affinities of carbon (assumed to be equivalent) act in the same plane and in mutually perpendicular directions, then from methane, CH_4, the following isomeric compounds arise on substituting the radicals R^1, R^2 and R^3 for hydrogen atoms—CH_3R^1 and CHR_3^1 have no isomers: $CH_2R_2^1$ and $CH_2R^1R^2$ each have two isomers: and $CH R^1R^2R^3$ has three isomeric forms, 'un nombre évidemment supérieur à celui que l'on connaît en réalité'.

(b) If however the four affinities are directed towards the apices of a regular tetrahedron with the carbon at its centre, CH_3R^1, $CH_2R^1_2$, $CH_2R^1R^2$ and $CH\ R^1(R^2)_2$ can exist in only one form, but there will be two forms of $CHR^1R^2R^3$ (or $CR^1R^2R^3R^4$). 'Si les affinités sont saturées par quatre groupes différents l'on obtient deux tétraèdres qui sont l'image non-superposable l'un de l'autre. Les composés $CR^1R^2R^3R^4$ présentent un cas différent de $C(R^1)_2R^2R^3$, $C(R^1)_3R_2$ ou $C(R^1)_4$.' Compounds $CR^1R^2R^3R^4$ may be said to contain an 'asymmetric carbon atom'.

2. *Relation between the 'asymmetric carbon atom' and optical activity*

(a) All compounds which are optically active in solution contain an asymmetric carbon atom (exemplified by lactic and tartaric acids, camphor, the sugars, etc.).

(b) Optical activity is lost when the asymmetry is removed.

(c) The converse of (a) is not necessarily true:

(i) the compound may consist of a mixture of equal quantities of each optical isomer;

(ii) not only the difference between the groups attached to the asymmetric atom but also their nature may have to be taken into account.

3. *The influence of the tetrahedral conception in unsaturated compounds, with two free affinities*

'La réprésentation de la liaison dite double entre atomes de carbone revient a deux tétraèdres ayant en commun une de leurs arêtes, tandis que les quatre groupes R^1, R^2, R^3, R^4, combinés occupent les sommets libres, situés dans un plan.'

When the groups R^1, R^2, R^3, R^4 are identical, or if $R^1 = R^2$ or $R^3 = R^4$, only one relative position is possible, but if R^1 and R^2 are not the same, and R^3 and R^4 are also unlike, irrespective of whether R^1 and R^3 or R^2 and R^4 are not identical, two isomers can exist (exemplified by the acids fumaric and maleic; bromo- and isobromo-maleic; citraconic and mesaconic; crotonic and isocrotonic).

4. 'La réprésentation de composés à liaison de carbone dite triple revient à deux tétraèdres qui ont en commun trois de leurs sommets, tandis que les groupes combinés occupent les deux sommets restés libre: il est clair que ce cas n'entraîne aucune prévision différente de celle des formules en usage.'

In comparing the contributions of Le Bel and van't Hoff it must be remembered that their stated objectives were not the same. Le Bel concerned himself solely with the relation between formulae and optical activity, while van't Hoff set out to establish a system of three-dimensional formulae, in which Le Bel's theme played an important, but not the only, part. In treating ethylenic derivatives, Le Bel, with characteristically Gallic logic, examines all the possibilities without committing himself, while van't Hoff, more sanguine and more imaginative, conceives a successful working model, which has not however proved fully consonant with modern theories. The two contrasted papers achieve a certain balance, for while in the one the isomerism of ethylenic compounds is fruitfully explained, in the other the existence of optically inactive and unresolvable meso-isomers of certain asymmetric compounds is made fully intelligible (see p. 96, (1), b).

From 1874 the principles of three-dimensional chemistry steadily rather than rapidly gained acceptance, but the study of molecular dissymmetry did not come into its own as an actively explored field of inquiry until the closing years of the century. In the 'theorem' concluding his paper (p. 97) Le Bel drew attention to the important fact that, since optical isomers are identical as regards the metric of their molecular structures, each must be formed in equal amounts from any synthetical process not subjected to any dissymmetric influence. Hence the preparation of a dissymmetrical compound must generally consist of two operations: first the synthesis of the d-l mixture, followed by its resolution into the optical isomers.

As Pasteur himself had shown, only one of all the numerous 'paratartrates' ultimately examined by him admitted of resolution by his crystallographic method of mechanically separating distinguishable crystals. He had, however, also demonstrated a second and in principle far more generally available method of resolution.* When paratartaric (racemic) acid is combined in aqueous solution with the dextro-base cinchonine (obtained from natural sources) the salt of pure l-tartaric acid is less soluble than the salt of the dextro-acid and therefore crystallizes first from the solution. The pairs of salts d-acid:l-base, l-acid:d-base or d-acid:d-base, l-acid:l-base do not differ in the metric of their structures and are equally soluble, but this is not the case for the pair d-acid:d-base and l-acid:d-base. But this method, so general in principle, often failed in practice, chiefly because naturally occurring acids and bases are only weakly functional, and their

* *Ann. Chim. (Phys.)* (3), **38**, 437 (1853).

salts are not fully stable in solution. For its full fruition therefore the method had to await the discovery of optically active strong acids and bases.

By their successful preparation in 1893* of camphor- and bromocamphor-sulphonic acids (from natural *d*-camphor) Kipping and Pope ushered in a period of intense activity, in which the possibility of molecular dissymmetry could be explored beyond the limits of carbon compounds. By 1902 it had been proved that dissymmetric compounds of tin,† sulphur,‡ and selenium§ could be prepared, and that the valencies of all these elements, when in their quadrivalent states, are tetrahedrally directed in space: a similar proof for silicon was delayed until 1904.‖ During the first years of the present century van't Hoff's original model for ethylenic compounds received its latest and most elegant proof. Carbon compounds of the types

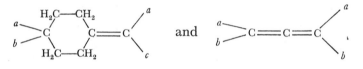

(Perkin, Pope and Wallach)¶   (Maitland and Mills)**

were proved to exist in optically active forms. In the original conceptions of Le Bel and van't Hoff the presence in a compound of an 'asymmetric carbon atom' must lead to dissymmetry, but the above structures show that the converse is not necessarily true, for neither contains an 'asymmetric carbon atom' in the original sense: rather the structures as a whole are dissymmetrical.

The spatial distribution of the valencies of nitrogen presented a far more protracted problem: indeed between the first suggestion offered by van't Hoff in 1878 and the final solution

* *J. chem. Soc.* **63**, 548 (1893).
† Pope and Peachey, *Proc. chem. Soc.* **16**, 42, 116 (1900).
‡ Pope and Peachey, *J. chem. Soc.* **77**, 1072 (1900); Smiles, *ibid.* 1174.
§ Pope and Neville, *J. chem. Soc.* **81**, 1557 (1902).
‖ F. S. Kipping, *Proc. chem. Soc.* **20**, 15 (1904); **23**, 9 (1907).
¶ *J. chem. Soc.* **95**, 1789 (1909).   ** *Nature, Lond.* **135**, 994 (1935).

during the present century almost 50 years elapsed, and the successes and vicissitudes of the various models proposed provide one of the most fascinating stories in the history of structural chemistry. The earliest observation bearing on the question was made by V. Meyer,* who hoped to contribute to a decision between the alternative views then held in regard to ammonium salts:

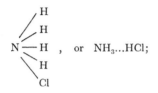

, or $NH_3...HCl$;

the latter formula shows the salt as a 'molecular compound', as required to satisfy Kekulé's dogma of constant tervalency for nitrogen. The same compound, of empirical formula $N(CH_3)_2(C_2H_5)_2I$, was found to result (1) from dimethylamine, $NH(CH_3)_2$, by the action of ethyl iodide, C_2H_5I, and (2) from diethylamine, $NH(C_2H_5)_2$, by the action of methyl iodide, CH_3I. Kekulé's formulation, which would be expected to necessitate the production of two isomeric salts

$$N(CH_3)_2(C_2H_5)...C_2H_5I \quad \text{and} \quad N(C_2H_5)_2(CH_3)...CH_3I$$

appeared therefore to be discredited.

Three years later van't Hoff proposed† for the derivatives of ammonium halides a cubical model (I)

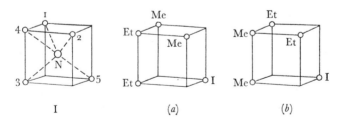

I (a) (b)

---

* *Ber. dtsch. chem. Ges.* **8**, 233 (1875); *Liebigs Ann.* **180**, 173 (1876).
† *Ansichten über die organische Chemie* (Braunschweig, 1878), p. 90.

wherein nitrogen occupies the centre, and the three positions 1, 2 and 3 are equivalent, while positions 4 and 5 'die supplementairen unter sich verschiedenen Valenzen bezeichnen'. The model therefore incorporates Kekulé's conception, and predicts at least two isomers *a* and *b* for V. Meyer's salts, upon the *identity* of which van't Hoff endeavoured to cast some doubt. Willgerodt, in 1888, suggested as a model a trigonal bi-pyramid*

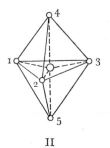

II

which resembled that of van't Hoff in distinguishing three equivalent (equatorial) directions (1, 2, 3) from the two collinear (polar) directions (4, 5).

Models of this type were by this time finding fewer adherents, and in 1890 Bischoff,† by proposing a tetragonal pyramid (III) as a model, broke away from Kekulé's views, and assigned

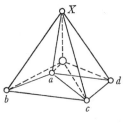

III

only one special position *X*, for the 'negative constituent' of the salt, the remaining *four* positions being all equivalent. The position of the nitrogen atom was supposed to be fixed by assuming the angle between the vertical direction and each of the four

* *J. prakt. Chem.* (2), **37**, 449 (1888).
† *Ber. dtsch. chem. Ges.* **23**, 1967 (1890).

other directions to be 120°. When the ammonium salt was dissociated and thereby lost the molecule, say Xd, it was supposed that the group b moved to occupy the now vacant position X, thus giving an amine molecule $Nabc$ in which all four units were co-planar, and the groups a, b, and c occupied the apices of a triangle with their directions separated by 120°.

In the same year such a planar model for the valency directions of *tervalent* nitrogen was challenged by Hantzsch and Werner,* who in a paper entitled 'Über räumlich Anordnung der Atome in stickstoffhaltigen Molecule', expressed the opinion 'die drei Valenzen des Stickstoffs bei gewissen Verbindungen nach der Ecken eines (jedenfalls nicht regulären) Tetraeders gerichtet sind, dessen vierte Ecke von Stickstoffatom selbst eingenommen wird'.† This proposition appeared to follow necessarily from the fact that a nitrogen atom can replace the group CH in acetylene

$$HC\equiv CH \qquad HC\equiv N \text{ (hydrocyanic acid)},$$

and was completely successful in explaining the various known isomeric forms of the oximes:

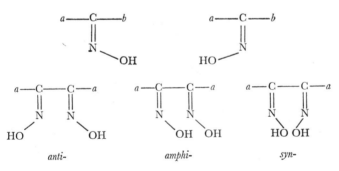

anti-	*amphi-*	*syn-*

In his exposition of the co-ordination theory in 1893 (see p. 121) Werner noted that since boron in the fluoro-borates RBF_4 and nitrogen in the ammonium salts NR_4X both manifest the co-ordination number 4, and for carbon the co-ordination number and the valency are both 4, it is probable that BF_4 and

* *Ber. dtsch. chem. Ges.* **23**, 11 (1890).
† In a note concluding the paper Hantzsch acknowledges that this conception 'in allem Wesentlichen des geistige Eigenthum des Hrn. Werner ist'.

NR_4 are tetrahedral radicals just as CH_4 is certainly a tetra-hedral molecule. Werner revived the question of the configura-tion of the ammonium radical NR_4 in 1902* and again advo-cated the tetrahedral model of his earlier conception.

A generation later, when a final solution was reached, much of the modern electronic theory of chemical union and valency was becoming clear, and in particular it had been shown how Werner's 'nebenvalenzen' could be assimilated to ordinary, primary valency. Further it was confirmed, also by the new theories, that the constitution of the ammonium ion, NH_4^+, was probably closely similar to that of methane, CH_4, as Werner had predicted. The two rival models of the ammonium radical NR_4 surviving in 1925 were both derived from earlier representations, shorn of the 'negative' (or ionizing) fifth valency:

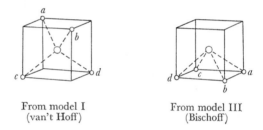

From model I   From model III
(van't Hoff)   (Bischoff)

Two researches, totally different in method, decided in favour of the tetrahedral model. The one testified once again to the pre-eminence of the property of dissymmetry to solve such problems, and the other was one of the earliest examples of the analysis of crystal structure by means of X-ray diffraction. Mills and Warren† were able to resolve into its optical isomers the sub-stituted ammonium bromide

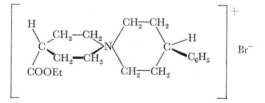

* *Liebigs Ann.* **322**, 261 (1902).   † *J. chem. Soc.* p. 2507 (1925).

of which the cation is stereochemically equivalent to

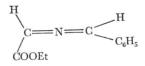

Its proved dissymmetry thus also proves that nitrogen in the reduced formula plays a role equivalent to that of the central carbon atom in the similar formula on p. 100. If its valencies were directed not tetrahedrally but according to Bischoff's model a plane of symmetry would contain the central nitrogen atom as well as the four end groups, and the compound prepared by Mills and Warren could not be dissymmetrical. Three years after this beautiful example of the classical method was completed, its results were confirmed by Wyckoff,* who analysed tetramethylammonium bromide, $N(CH_3)_4Br$ by X-ray diffraction, and found the cation to be in the form of a regular tetrahedron, with central nitrogen.

The stereochemical models of Willgerodt (II) and Bischoff (III) have by no means disappeared from the field of stereochemistry: in phosphorus pentafluoride, PF_5, in vaporized phosphorus pentachloride, PCl_5 and in arsenic pentafluoride, AsF_5, the five valencies of the central atom are directed as in model II, and in iodine pentafluoride, IF_5, as in Bischoff's model (III).

BIOGRAPHICAL AND OTHER REFERENCES

Chemical Society Memorial Lectures:
 L. Pasteur, *J. chem. Soc.* **71**, 683 (1897).
 J. Wislicenus, *J. chem. Soc.* **87**, 501 (1905).
 J. H. van't Hoff, *J. chem. Soc.* **103**, 1127 (1913).
On Molecular Asymmetry. L. Pasteur, Alembic Club (translated by Alex. Scott). (1897.)
Titles and chronology of J. H. van't Hoff's contributions to stereochemical theory:
 (1) *Voorstel tot Uitbreidung der Structuur-Formules in de Ruimte.* (Amsterdam, 1874.)
 (2) *La chimie dans l'espace.* (Rotterdam, 1875.)

* *Z. Kristallogr.* **67**, 91 (1928).

(3) *Die Lagerung der Atome im Raume.* Foreword by J. Wislicenus. (Vieweg, Braunschweig, 1877). (Based on the French text (2), with much amplification by the translator and by van't Hoff.)

(4) *Ansichten über die organische Chemie.* (Braunschweig, 1878.)

(5) *Dix Années dans l'Histoire d'une Théorie.* Second Edition of (2), with dedication 'à M. J-A. Le Bel en témoinage de ma respectueuse affection'. (Rotterdam, 1889.) (In this publication the full French texts of Le Bel's paper of 1874, and of van't Hoff's brochure (2) are reprinted, together with the German text of the contemporary comments on (2) by Wislicenus and by Kolbe.)

(6) *The arrangement of Atoms in Space.* (London, 1898.) (van't Hoff appears to have overlooked that (5) was called by him 'Second Edition', and the German text issued from Amsterdam in 1894, of which (6) is a translation by Eiloart, is sub-titled 'Second Revised and Enlarged Edition'. It contains a new Preface by Wislicenus, dated from Leipzig, 1894, and an Appendix by Werner.)

6

ARRHENIUS

Ions and the concept of electrovalency

Wie sich damals z.b. der Druck des Chlorammoniumdampfes bei Zugrundelegung (Avogadros) Gesetzes zu gross ergab, so ist in mehreren Fällen der osmotische Druck ein abnorm grosser: Abweichung dieser Art sich bei der Mehrzahl der Salze zeigen, deren Spaltung in gewöhnlicher Weise nur schwer anzunehmen ist.... Demnach erscheint es gewagt ein Avogadrosches Gesetz fur Lösungen derart in Vordergrund zu stellen und ich würde mich dazu auch nicht entschlossen haben, hätte nicht Arrhenius mich brieflich auf die Wahrscheinlichkeit hingewiesen, dass es sich bei Salzen um ein Spaltung in Jonen handelt.   VAN'T HOFF, *Z. phys. Chem.* **1**, 500 (1887)

(*Just as, according to Avogadro's law, the vapour pressure of ammonium chloride proved too great, so in many cases osmotic pressures are abnormally large: exceptions of this sort are shown by most salts, the dissociation of which in any ordinary sense is very difficult to admit. Thus it would seem rash to set the application of Avogadro's law to solutions as strongly in the foreground as I have done, and I would not have decided upon this, had not Arrhenius, in a letter, pointed out the probability that for salts and similar substances it is a question of dissociation into ions.*)

By the year 1900 three great generalizations had come to light, each of which exerted a revolutionary influence upon physical science—the atomic theory, the kinetic theory of matter, and the quantum theory: among these the ionic theory may take an equal rank. After nearly a century of atoms the theory added ions to the chemist's repertory of the constituent units of chemical compounds. It formed the indispensable basis of Werner's exploration of a vast array of 'molecular compounds' during the period 1893–1913, which led ultimately to a great extension of the previous conceptions of valency. Under its influence the old dualistic theory underwent a metamorphosis into a practicable and viable form, which brought the electrical nature of chemical union finally and permanently into the forefront of chemical theory.

Svante August Arrhenius (1859-1927)

As a young student at the University of Upsala Arrhenius owed much to the good offices of Ostwald, then professor at Riga. A doctoral thesis upon the electrical conductivity of dissolved electrolytes presented in 1883 had secured only the rank 'non sine laude approbatur' and Arrhenius had almost abandoned the hope of an academic position, when Ostwald, during a visit to Upsala, in no way concealing his discernment, made him the offer of the post of Privatdocent at Riga. To this the countrymen of Berzelius could not remain indifferent, with the result that Arrhenius was nominated to a valuable 'Reise-stipendium' in the control of the Swedish Academy of Sciences. So supported he was enabled to spend nearly three 'Wanderjahre' in the company of those whose scientific interests most converged upon his own—first with Ostwald at Riga, then with Kohlrausch and Boltzmann at Würzburg and Graz, and with van't Hoff at Amsterdam. It was during such fortunate encounters that the ionic theory must have assumed its definite form: and the first volume of the *Zeitschrift für physikalische Chemie*, issued under the joint management of van't Hoff and Ostwald, contained the title 'Sv. Arrhenius, Über die Dissociation der in Wasser gelösten Stoffe', dated 27 December 1887. Six years later Arrhenius was honoured by the Nobel Prize for Chemistry.

The antecedents of the theory finally announced by Arrhenius can be traced back to a theoretical discussion published by J. L. Gay-Lussac in 1839, under the title 'Considerations sur les forces chimiques'.

...au moment du mélange (des sels dissous) il y a un véritable pêle-mêle entre les acides† et les bases,† c'est-a-dire que les acides se combinent indifféremment avec les bases et réciproquement: peu importe l'ordre de combinaison, pourvu que l'acidité et l'alcalinité soient satisfaites et bien évidemment elles le sont, quelque permutation qui s'établisse entre les acides et les bases.

Ce principe d'indifférence de permutation (*d'équipollence*) établi, les

* *Ann. Chim. (Phys.)*, **70**, 407 (1839).
† In 1839 'acide' meant *acidic oxide*, and 'base' *basic oxide*.

décompositions produites par double affinité s'expliquent avec une très heureuse simplicité. Au moment du mélange de deux sels neutres il s'en forme deux nouveaux dans des rapports quelconques avec les deux premiers; et maintenant, suivant que l'insolubilité sera plus prononcée pour les nouveaux sels que pour les sels donnés, il y aura trouble d'équilibre et séparation d'un sel, quelquefois même de plusieurs.

A year later H. Hess provided a quantitative basis to Gay-Lussac's theory:*

Nimmt man zwei Lösungen neutraler Salze, die gleiche Temperatur haben, und durch gegenseitige Zerzetzung zwei neue Salze erzeugen, so ändert sich die Temperatur nicht....So dass neutrale zusammengemengte Lösungen *thermoneutrale* sind.

(*When two solutions of neutral salts, each at the same temperature, are mixed and by double decomposition two new salts are generated, there is no change of temperature: ...so that neutral (salt) solutions are* thermoneutral *on being mixed.*)

Hess deduced this theorem of thermoneutrality from his previous determinations of the heats of neutralization of acids by bases. If for the heat of neutralization in dilute aqueous solution of one equivalent of acid A by the base B we use the symbol $H[A/B]$, Hess proved that

H [sulphuric acid/potash] + H [nitric acid/lime] =

H [nitric acid/potash] + H [sulphuric acid/lime].

A. W. Williamson, in 1851,† explained the formation of di-ethyl ether by the interaction of sulphuric acid and ethyl alcohol as due to a sequence of double decompositions which he regarded as analogous to the interaction between hydrochloric acid and silver nitrate, leading to the separation of silver chloride.

In the first stage C_2H_5OH and H_2SO_4 to some extent exchange radicals so that $H.OH$ and $C_2H_5.H.SO_4$ are formed: in the latter stage $C_2H_5.H.SO_4$ and $C_2H_5O.H$ exchange radicals, and di-ethyl ether results because its volatility ensures its continuous removal from the liquid mixture. In a similar way an atom of hydrogen in a solution of hydrochloric acid does not always remain bound to the same chlorine atom, but exchanges it for new atoms of chlorine, one after another. If solutions of hydrochloric acid and silver nitrate are mixed some molecules of HNO_3 and $AgCl$ are immediately

* 'Thermochemische Untersuchungen', *Poggendorf's Ann.* **52**, 107 (1841).
† *Ann. Chim. (Pharm.)*, **77**, 43 (1851).

formed. Being only slightly soluble the molecules of AgCl separate from solution, so that HCl and AgNO$_3$ are not reformed. The process continues until there remains in solution only the small amount of AgCl needed to saturate it. The insolubility of AgCl and the volatility of ether cause the chemical actions producing them to move only in one direction.

R. Clausius* considering the conduction of electricity by dissolved electrolytes was more cautious:

Williamson speaks of perpetual exchange of hydrogen atoms between HCl molecules, whereas for the conduction of electricity it suffices that at the collisions of the molecules now and then, and perhaps relatively seldom, an exchange of the partial molecules takes place.

C. A. Valson† undertook a study of the capillary properties of salt solutions. He found that among salt solutions of equivalent (normal) concentration that of ammonium chloride exhibited the highest capillary rise, and consequently chose this solution as reference for others.

...les *modules capillaires* expriment les abaissements de hauteur capillaire qui résultent de la substitution d'un radical métallique à l'ammonium ou d'un radical metalloïdique au chlore. Ces mêmes nombres jouissent encore d'une autre propriété importante qu'on peut énoncer ainsi:

Le module d'un sel est égal à la somme des modules de ses deux radicaux: par exemple,

Module pour ½ Ba,   3·9 mm.
Module pour NO$_3$,   1·0 „
Somme pour ½ Ba(NO$_3$)$_2$, 4·9 „
NH$_4$Cl, hauteur du sol. normale 60·9 mm.
Ba(NO$_3$)$_2$ sol. normale:
    calculé   56·0 „
    observé   55·9 „

F. W. Kohlrausch was the first to devote himself to the accurate measurement of the electrical conductivity of solutions of electrolytes, under varying conditions of concentration and temperature. The extensive data were published in 1879 under the title 'Das electrische Leitungsvermögen der wasserigen Lösungen von Hydraten und Salzen den leichten Metalle, sowie von Kupfervitriol, Zinkvitriol, und Silbersalpeter'.‡

---

* *Poggendorf's Ann.* **101**, 338 (1857).
† *Ann. Chim.* (*Phys.*), (4), **20**, 361 (1870).
‡ Wiedermann's *Ann. Phys., Lpz.*, **6**, 145–210 (1879).

Je mehr die Anzahl der Wassertheilchen diejenige des Electrolytes überwiegt, desto mehr wird wesentlich nur die moleculäre Reibung der Jonen an den Wassertheilchen, nicht aber ihre Reibung aneinander, in Betracht kommen. . . .

Hiernach muss also jedem electrochemische Elemente—z.b. H, K, Ag, NH_4, Cl, J, NO_3, $C_2H_3O_2$—in verdünnte wasseriger Lösung ein ganz bestimmter Widerstand zukommen, gleichgültig aus welchem Electrolyte der Bestandtheil abgeschieden wird. Aus diesen Widerstanden, welche für jedes Element ein für allemal bestimmbar sein mussen, wird sich das Leitungsvermögen jedes verdünnter Lösung berechnen lassen.

(*The more the number of water particles exceeds those of the electrolyte the more the friction between the ions themselves gives place to friction only between ions and water particles. . . . Consequently each electrochemical element in dilute aqueous solution suffers a constant resistance, no matter from what electrolyte it was dissociated. From the values of such resistance, determinable once for all for each element, the conductivity of every dilute (salt) solution may be calculated.*)

F. M. Raoult, who in a study, extending over many years, of the freezing points of aqueous and non-aqueous solutions, laid the foundations of the cryoscopic method of determining molecular weights, published in 1885 a concluding paper 'Sur le point de congélation des dissolutions salines'.*

[From the section on 'Constitution spéciale des sels dissous dans l'eau'.]
. . .la diminution des hauteurs capillaires, l'accroissement des densités, la contraction du protoplasm, l'abaissement du point de congélation, bref, la plupart des effets physiques produit par les sels sur l'eau dissolvante sont la somme des effets produits séparément par les radicaux électropositifs et électronégatifs qui les constituent, et qui agissent comme s'ils étaient simplement mélangés dans la liquide.

Ce fait, sans être une conséquence nécessaire de la théorie dualistique et électrochimique des sels, la confirme cependant dans son principe. Il montre que les sels en dissolution dans l'eau doivent être considérée comme des systèmes de particules, dont chacune est formée d'atomes solidaires, et garde, *malgré son état de combinaison avec les autres*, une grande partie de son individualité, de son action et des ses caractères propres.

[On molecular depressions of freezing point by acids and their salts]

HCy	19·4	KCy	32·2
$H_2C_4O_2$	19·0	$KC_2H_3O_2$	34·5
$H_2C_2O_4$	23·2	$K_2C_2O_4$	45·0

Ces différences énormes résultent de ce que les acides faibles présentent toujours un abaissement moléculaire anormal, c'est-à-dire trop faible de près de moitié, comme si la plupart de leurs molécules étaient soudées deux à deux, tandis que leurs sels alcalins ne le font jamais.

* *Ann. Chim. (Phys.)*, (6), **4**, 401 (1885).

In 1911 Arrhenius, by then Director of the Physico-chemical Department of the Nobel Institute of the Swedish Academy of Sciences at Stockholm, delivered at Yale University, upon the Silliman Foundation, a course of eleven lectures, published in 1912* under the title *Theories of Solution*. In Lecture XI on 'the Theory of Electrolytic Dissociation' he gave an interesting account of the immediate antecedents of his final paper of 1887.

In 1883 I investigated the conductivity of electrolytes as dependent on their concentration and temperature† and concluded that their solutions contain two different kinds of molecules, of which one is a non-conductor, the other conducting electricity in consequence of properties attributed to it by the hypotheses of Gay-Lussac, Williamson and Clausius. These latter were called simply *active* molecules, and their number, increasing with dilution at the expense of the inactive molecules, tends to a limit, probably first reached when all inactive have been transformed into active molecules. At very high dilutions the additive property postulated by Kohlrausch is not only true within certain groups of similar electrolytes but for all electrolytes of whatsoever composition. An acid is the stronger the greater its conductivity, and at infinite dilution all acids have equal strength....The results of Ostwald's measurements of chemical equilibria were discussed, and explained as dependent on the lowering of the activity of weak acids caused by the presence of their salts and of strong acids. The heat released in the neutralization of a wholly active acid by a wholly active base is always the same, and equal to the heat consumed in the activation of an equivalent quantity of water. The deviation of weak acids and bases from the constant value is due to the heat needed for their activation....When this memoir was written the measurements of Raoult on the freezing points of saline solutions had not appeared. Therefore it was regarded as too bold to assert that the active molecules were dissociated into their ions, and it was only maintained that they should be subject to the hypothesis of equipollency.‡

Immediately after the publication of Raoult's memoir in 1885 I calculated the coefficient of activity from the conductivity data and the degree of dissociation which was necessary to explain the values of the van't Hoff coefficient *i* calculated from Raoult's data. A very good agreement was found and the basis for an open declaration of the state of dissociation of electrolytes was found strong enough: the word activity was replaced by the term *electrolytic dissociation*.

In his paper of 1887 Arrhenius rested his case mainly upon agreements which were at best semi-quantitative, and he was

---

* Yale University Press.
† *Bih. svensk. VetensAkad. Handl.* **8**, 13, 14; in French (Stockholm, 1884).
‡ Gay-Lussac, 1839 (p. 110).

obliged to admit numerous unexplainable exceptions. It was again the penetrating vision of Ostwald that perceived the possibility of a more cogent proof of the hypothesis of electrolytic dissociation. The laws of thermodynamics, being independent, stand in a certain sense as arbiters of chemical theories, and Ostwald proposed to submit Arrhenius' theory of equilibrium between molecules and ions to the test of the law of mobile equilibrium, which is derived directly from thermodynamic principles. The results, now long known in the form of 'Ostwald's dilution law', proved for the case of weak acids to be at that time probably the most exact example of obedience to the thermodynamic law. On the other hand, Planck, for the class of strong electrolytes—salts in general, as well as the strong acids and bases—found the supposed dissociation completely incompatible with the law. Since those not disposed to credit the theory of ionic dissociation naturally felt that the behaviour of the highly dissociated electrolytes ought to be crucial, their hostility appeared well founded and actively survived into the present century: it was not finally overcome until the modern study of crystal structure had proved that solid strong electrolytes are themselves composed of ions already in being, and that no *chemical* equilibrium exists in their solutions.

BIOGRAPHICAL AND OTHER REFERENCES

Chemical Society Memorial Lecture:
    Sv. Arrhenius, *J. chem. Soc.* 1928, 1380.
Theories of Solution. S. Arrhenius. (Oxford University Press for Yale University Press, London, 1912.) (Silliman Memorial Lectures delivered at Yale University, 1911.)

7

WERNER

A new type of valency

Immer überzeugender sprechen die Tatsachen der anorganische Chemie dafür, dass unsere aus den Konstitutionsverhältnissen der Kohlenstoffverbindungen entwickelten Valenzvorstellungen kein genügendes Bild vom Molekülbau der anorganische Verbindungen abzuleiten gestatten. Infolgedessen mehren sich die Bestrebungen, durch Erweiterung der Valenzlehre eine breitere theoretische Grundlage für die Lehre von der Konstitution der anorganische Verbindungen zu gewinnen....Dabei bleiben wir uns wohl bewüsst, das unsere neuen Vorstellungen nur zusammenfassende Bilder sind, von denen wir nicht mehr als von jener anderen naturwissenschaftlichen Hypothese erwarten konnen, nämlich nur, dass sie uns eine Ökonomie des Denkens und Beschreibens ermöglichen. Mögen diese Bilder später durch andere, vollkommenere abgelöst werden, so dürften sie doch das Verdienst gehabt haben, die Systematik des fast unübersehbar Tatsachenmaterials der anorganische Chemie angebahnt zu haben.

ALFRED WERNER, from the preface to *Neuere Anschauungen auf dem Gebiete der anorganische Chemie* (1905)

(*The facts of inorganic chemistry continually convince us that from our valency theory, developed from the constitution of organic compounds, no adequate representation of the molecular structure of inorganic compounds can be derived. Consequently efforts are increasingly made to broaden the scope of that theory and so to obtain a wider basis for the laws of the constitution of inorganic compounds....However we remain aware that our new notions provide only pictures in outline, from which we can expect, as from all scientific hypotheses, no more than economy in thought and description. It may be that our present ideas will serve in the future to open ways through the vast extent of the factual material of inorganic chemistry towards a fuller understanding.*)

Alfred Werner (1866-1919)

A native of Mülhausen in Alsace, Werner at the age of nineteen entered the Technische Hochschule at Karlsruhe, but moved a year later to the Polytechnikum at Zürich where instruction in chemistry was under the direction of Lunge and Hantzsch.

After graduating in 1890 at the University of Zürich with a dissertation 'Über die räumliche Anordnung der Atome in stickstoffhaltige Molekülen' (cf. p. 103), he spent a winter in Paris under Berthelot, and in 1892 gained his master's degree at the Zürich Polytechnikum with an exposition 'Beiträge zum Theorie der Affinität und Valenz' which already outlined the ideas round which he was to build his life's work. Less than a year later came the now famous 'Beitrag zur Konstitution anorganischer Verbindungen'* and a call to the chair of chemistry in the University of Zürich.

Few chemists can have been so well inspired that after a quarter of a century's rigorous testing in the laboratory hardly a detail needed modification in a programme laid down before the experimental work began. In spite of many pressing invitations to other chairs Werner remained faithful to Zürich until his untimely death in 1919 at the early age of 53 years, six years after he had been awarded the Nobel Prize for Chemistry.

It was by no means without precedent that Werner's introduction in 1893 of what appeared to be a new type of affinity should have been little regarded by the majority of chemists. The ionic theory, which from six years before had steadily penetrated into the field of inorganic chemistry in the remaining years of the century, was heralded by a long series of past tentatives in its own direction, but Werner's theory possessed in the eyes of many conservative chemists the doubtful merit of complete novelty. It was not until 1911, when a dissymmetric 'co-ordination compound' was prepared and resolved into its optical isomers, that such a link with well-recognized and familiar principles recommended the new ideas to general notice. Had Werner's life extended to the common span he could have witnessed the establishment of the co-ordination theory as a fundamental tenet of the modern electronic theory of valency. The work on dissymmetrical co-ordination complexes of cobalt, carried out at Zürich during the year 1911, was published in a series of five papers, under the general title 'Zur Kenntnis des asymmetrischen Kobaltatoms' (p. 121).

* *Z. anorg. Chem.* **3**, 267 (1893).

The following is a synopsis of the essay by Werner:

Beitrag zur Konstitution anorganischer Verbindungen*

(1) Metal-ammonia compounds are formed from ammonia and metallic salts by reactions similar to that between ammonia and hydrochloric acid (regarded as a halide salt of hydrogen) by which ammonium chloride is produced. In the past the classification of these compounds was arbitrary in that the most stable were represented by the usual type of constitutional formulae while the less stable were relegated to the class of 'molecular compounds'. A system more in accord with present knowledge rests upon their empirical formulae and certain of their properties:

Class I—Compounds containing *six* ammonia molecules to one metal atom: e.g. $Co6NH_3Cl_3$

Class II—Compounds containing *four* ammonia molecules to one metal atom: e.g. $Cu4NH_3(NO_3)_2$.

Each of these classes may be subdivided according to the valencies of the metals:

$$Pt(NH_3)_6Cl_4: \quad Co(NH_3)_6Cl_3: \quad Ni(NH_3)_6Cl_2.$$

Ammonia can be replaced by other groups or molecules, for example water, but *the total number of groups remains constant at six*—

$$Cr(NH_3)_6Cl_3: \quad Cr(NH_3)_5H_2OCl_3: \quad Ni(NH_3)_3(H_2O)_3SO_4\ldots.$$

Metal-ammonia compounds containing the complexes MA_6 or MA_4 may by loss of groups A give rise to derivatives containing fewer groups A—

$$Co(NH_3)_6Cl_3 \longrightarrow \begin{cases} Co(NH_3)_5Cl_3 \\ Co(NH_3)_4Cl_3 \\ Co(NH_3)_3Cl_3 \end{cases}$$

$$Pt(NH_3)_4Cl_2 \longrightarrow \begin{cases} Pt(NH_3)_3Cl_2 \\ Pt(NH_3)_2Cl_2 \end{cases}$$

(2) *Blomstrand* in 1869 attempted to formulate metal-ammonia compounds by supposing them to contain nitrogen chains analogous to those of carbon—

for $Pt(NH_3)_4Cl_2$ $\quad Pt\begin{cases} NH_3-NH_3-Cl \\ NH_3-NH_3-Cl \end{cases}$

and *Jörgensen*, by an extended experimental study of these compounds, sought to support Blomstrand's formulation.

A. W. Hofmann, and others, viewed the complex compounds as ammonium salts in which hydrogen is replaced partly by metal atoms and partly by the ammonium radical NH_4

* *Z. anorg. Chem.* **3**, 267–330 (1893).

A NEW TYPE OF VALENCY

for $Pt(NH_3)_4Cl_2$,

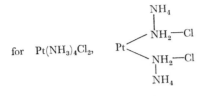

Hofmann's formulation was however disproved when Jörgensen found that tertiary amines, NR_3, phosphines, PR_3, and pyridine, C_5H_5N, could replace ammonia.

(3) From a solution of the compound $Cr(NH_3)_6Cl_3$ silver nitrate precipitates all the chlorine atoms as silver chloride. This compound loses one molecule of ammonia in the direct reaction

$$Cr(NH_3)_6Cl_3 = NH_3 + Cr(NH_3)_5Cl_3,$$

but from a solution of the product silver nitrate precipitates only *two* chlorine atoms as silver chloride: and by the action of concentrated sulphuric acid only two atoms of chlorine are released, as two molecules of hydrogen chloride, the third being unaffected.

The loss of one molecule of ammonia is accompanied by a functional change of one of the acid radicals, which no longer plays the part of an ion. Only two chlorine atoms act as ions, while the third behaves as an analogue of chlorine in chloroethane.

(Mit diesem Verlust eines Ammoniakmoleküls tritt aber gleichzeitig Funktionswechsel eines Saurerestes ein. Während sich der betreffende negative Komplex vor dem Austritt des Ammoniakmoleküls als Jon verhielt, hat er nach demselben die Eigenschaft, als Jon zu wirken, eingebüsst.... Zwei Chloratome verhalten sich as Jonen während das dritte sich ganz analog verhalt wie Chlor in Chloräthan.)

Other examples are

$$\left(Co\,\frac{(NH_3)_5}{X}\right)X_2, \quad \left(Ir\,\frac{(NH_3)_5}{X}\right)X_2, \quad \left(Rh\,\frac{(NH_3)_5}{X}\right)X_2.$$

The question now arises, how many times may this process be repeated?

(4) To compounds of the type $M(NH_3)_6Cl_3$ and $M(NH_3)_5Cl_3$ Jörgensen ascribed the formulae

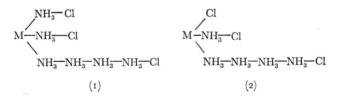

(1)   (2)

For further successive loss of one molecule of ammonia we should have

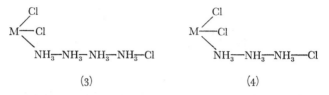

$$(3) \qquad\qquad (4)$$

Since in (1) and (2) the chlorine atom functions as an ion whether it be bound to the metal by one or by four NH_3 groups it must also so function in (4) where it is bound by three such groups. But it is found that in all compounds of the type $M(NH_3)_3X_3$ *none* of the three negative radicals behaves as an ion. From the compound $Ir(NH_3)_3Cl_3$ concentrated sulphuric acid, even on heating, expels no hydrogen chloride, while from $Ir(NH_3)_4Cl_3$ and $Ir(NH_3)_5Cl_3$ one and two molecules of HCl are expelled respectively.

A further loss of ammonia from (4) above would yield a substance $M(NH_3)_2Cl_3$, but no metal-ammonia compound of such a type has ever been prepared. *The loss of a fourth molecule of ammonia entails its replacement by a fourth acid radical to give the type* $M(NH_3)_2X_4$.

The successive functional changes operating in all radicals MR_6 is most prominently seen in the cobalt series—

$$(Co(NH_3)_6)X_3, \quad (Co(NH_3)_5(NO_2))X_2, \quad (Co(NH_3)_4(NO_2)_2)X,$$
$$(Co(NH_3)_3(NO_2)_3), \quad (Co(NH_3)_2(NO_2)_4)K.$$

The next number, $(Co(NH_3)(NO_2)_5)K_2$ is unknown, so it is all the more striking that the series ends with $(Co(NO_2)_6)K_3$, well-known as potassium cobaltinitrite. Such a characteristic transition from the basic metal-ammonia radicals to others of parallel constitution which are acidic is impossible to explain by the Blomstrand–Jörgensen conception, which therefore appears to me to be untenable.

(Diesen eigenthümlichen Übergang von den basischen Metallammoniak-radikalen zu ähnlichen als Säure wirkenden Komplexen vermag die Blomstrand–Jörgensensche Auffassung in keiner Weise zu erklären, und dieselbe erscheint mir deshalb unhaltbar.)

Examples of salts containing the acid radical (MX_6) are very numerous:

M tervalent: Cyanides, $R_3M(Cy)_6$—

M = Co, Rh, Cr, Mn, Ir, Fe.

Chlorides, $R_3M(Cl)_6$—

M = Cr, Fe, Ir, Rh.

Fluorides, $R_3M(F)_6$—

M = Al, Fe, Cr.

Nitrites, $R_3M(NO_2)_6$—

M = Co, Rh.

A NEW TYPE OF VALENCY

M quadrivalent: Halides, $R_2M(Hal)_6$

$M = Pt, Pb, Si, Sn, Ti, Zr.$

M bivalent: Cyanides, $R_4M(Cy)_6$

$M = Fe, Ru, Os, Mn.$

(5) *Salt hydrates.* By exchanging all molecules of ammonia in a metal-ammonia complex MA_n for water molecules the commonest forms of the hydrates of salts of the metal M result. (Werner here lists 50 salts the highest hydrates of which contain $6H_2O$ per formula-weight.) The very common occurrence of hexahydrates and of metal-ammonia complexes containing 6 molecules of ammonia independently of the acids of the salts, and of metals in double-salts united to six acid radicals, e.g. $Fe(OH_2)_6 . Cl_3$, $Fe(Cy)_6 . K_3$, $FeF_6 . K_3$ cannot be a matter of accident, but takes its innermost cause from the metal atom.

(...schliessen wir dass die Eigenschaft der Metallsalzen sehr haufig sechs Wassermoleküle bezw. sechs Ammoniakmoleküle zu binden, dieselbe dem Metallatom innewohnende Ursache hat.)

Electrolytic dissociation of salts. The first condition for the electrolytic dissociation of a salt is the power of its metal atom to bind a definite number of water molecules into a radical in which we must suppose the water molecules so arranged that a direct connection between metal atom and the acid radical can no longer exist. Other solvents can bring about such dissociation only if they can form similar radicals with metal salts.

(Die erste Bedingung zum elektrolytischen Dissoziation eines Salz ist die Fähigkeit seines Metallatoms sich mit einer bestimmten Anzahl von Wassermolekülen zu einem Radikal zu verbinden, in dem wir uns die Wassermoleküle so angeordnet zu denken haben, dass eine direkte Bindung zwischen dem Metallatom und dem Saurerest nicht mehr eintreten kann.... Nur solche Lösungsmittel sind befähigt, elektrolytische Dissoziation zu bedingen welche mit den Metallsalzen zu analogen Radikalen zusammentreten können.)

(6) *Distinction between the mode of action within the complex of molecules such as H_2O, NH_3, PR_3, and that of univalent radicals.* The valency of the complex MA_6 is equal to the difference between the valency of the metal and the total valencies of the univalent groups within the complex, independently of molecules such as H_2O, NH_3, etc. which are present in the complex:

$Co(NH_3)_6$ $3 - 0$ $Co(NH_3)_6 . X_3$

$Co\binom{(NH_3)_3}{(NO_2)_3}$ $3 - 3$ (No acid radical)

$Co\binom{(NH_3)_2}{(NO_2)_4}$ $3 - 4$ $Co\frac{(NH_3)_2}{(NO_2)_4} . K$

$Fe(CN)_6$ $2 - 6$ $Fe(CN)_6 . K_4$

$Pt(Cl)_6$ $4 - 6$ $Pt(Cl)_6 . K_2$

The mode of action of NH_3, H_2O, etc. is specific and peculiar: these molecules are able to transmit the affinity forces of the metal outwards from the complex. Imagine the metal ion to be a sphere with positive charge on its surface: an envelope of water molecules bound to it will be charged negatively on its inside and therefore positively on the outside of the complex. When the complex is brought into aqueous solution more water molecules will be attracted and bound to the complex in the same way as the original six molecules. As a consequence the positive and negative complexes of salt will be effectively separated, and electrolytic dissociation takes place: 'diese Dissoziation durch die Eigenschaft des Wassers als Kraftübertrager zu wirken bedingt wird'.

(7) *The principle of co-ordination—co-ordination number.* 'The number of atomic groups which can become united to an atom in the ways already described I term the *co-ordination number* of the given atom: an atom with co-ordination number 6 has the property of directly binding or *co-ordinating* six groups. The position occupied by one such group I term a *co-ordination site.*'

In the first series of the Periodic System boron, carbon and nitrogen all have co-ordination numbers 4, but only for carbon is this equal to the valency number. Boron, on the positive side of carbon, adds to its compounds with negative univalent elements one univalent element, and the resulting complex behaves as a univalent radical, for example $(BF_4).M$. Nitrogen, negative in regard to carbon, adds to its compounds with positive radicals one more univalent group, resulting in the formation of a positive univalent radical, as in the ammonium compounds $NR_4.X$. Thus the valency numbers of boron and nitrogen are 3 but their co-ordination numbers are 4.

For atoms in general co-ordination number and valency number are unequal: the *valency number* is the maximum number of univalent atoms which *in the absence of other atoms* combine directly with a given atom: the *co-ordination number* is the maximum number of atoms or groups with which a given atom can combine directly.

Two molecules which each have one co-ordination site vacant can combine—$B(CH_3)_3$, trimethyl boron, and ammonia yield the compound $(CH_3)_3B.NH_3$, a solid crystalline substance with melting-point 56° which can be distilled without decomposition: platinic chloride, and trimethylstannic iodide, each with *two* vacant sites yield respectively

$$\left(Pt \, \frac{(NH_3)_2}{Cl_4}\right) \quad \text{and} \quad \left(SnI \, \frac{(C_2H_5)_3}{(NH_3)_2}\right):$$

Iridium trichloride, cobaltic nitrite with *three* vacant sites form

$$Ir \left(\frac{(NH_3)_3}{Cl_3}\right) \quad \text{and} \quad Co \left(\frac{(NH_3)_3}{(NO_2)_3}\right):$$

The radical CoX_2 has *four* sites vacant, but here the valency of the complex

$$Co \left(\frac{(NH_3)_4}{X_2}\right)$$

must be satisfied outside the complex, to give

$$\mathrm{Co}\left(\begin{array}{c}(\mathrm{NH_3})_4\\X_2\end{array}\right).X:$$

Similarly the radical CoX has *five* sites, and yields

$$\mathrm{Co}\left(\begin{array}{c}(\mathrm{NH_3})_5\\X\end{array}\right).X_2,$$

and the final stage is $\mathrm{Co(NH_3)_6}.X_3$.

(8) *The configuration of co-ordination complexes.* (*a*) Radicals MR_6: if the metal atom is considered to lie at the centre of a regular octahedron, the apices of which are occupied by the radicals, then there should be two isomeric forms of the complexes

$$\left(M\begin{array}{c}A_4\\X_2\end{array}\right) \quad \text{or} \quad \left(M\begin{array}{c}A_2\\X_4\end{array}\right),$$

and such isomers were proved by Jörgensen to exist among the complexes of both cobalt and platinum.

(*b*) Radicals MR_4: the configuration of this complex may be derived from the octahedral MR_6 by the removal of two sites. The removal of two *polar* sites leaves the metal atom and the four attached groups lying in the same plane, an arrangement by which the two known isomeric forms of the platinous compounds $\left(\mathrm{Pt}\begin{array}{c}(\mathrm{NH_3})_2\\X_2\end{array}\right)$ are readily explained as

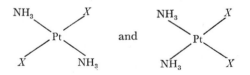

The existence of these isomeric forms proves that the arrangement cannot be tetrahedral.

On the other hand it may well be assumed that in the boron radical BF_4 and the ammonium radical NR_4 the four groups occupy the apices of a regular tetrahedron, as they certainly do in carbon compounds.

The following is a summary of the papers entitled:

Zur Kenntnis des asymmetrischen Kobaltatoms

(I) *Ber. dtsch. chem. Ges.* **44**, 1887 (1911):

It had been previously found that a molecule of ethylene-diamine, $\mathrm{NH_2CH_2CH_2NH_2}$ (symbol *en*) could stably occupy two co-ordination sites, and thus be substituted for two molecules of ammonia.

The complexes in the salts

$$\begin{bmatrix} Cl \\ NH_3 \end{bmatrix} Co(en)_2 \end{bmatrix} X_2 \quad \text{and} \quad \begin{bmatrix} Br \\ NH_3 \end{bmatrix} Co(en)_2 \end{bmatrix} X_2$$

are dissymmetric, if the octahedral model is assumed, as in the spatial formulation—

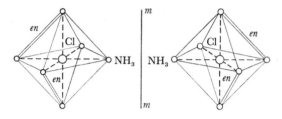

(mm = mirror plane)

and the optical isomers were successfully separated; and found to be unusually stable towards racemization. Activity was retained even after substituting a water molecule for the halogen atom by means of the reaction

$$\begin{bmatrix} Br \\ NH_3 \end{bmatrix} Co(en)_2 \end{bmatrix} Br_2 + 3AgNO_3 + H_2O = \begin{bmatrix} H_2O \\ NH_3 \end{bmatrix} Co(en)_2 \end{bmatrix} (NO_3)_3 + 3AgBr.$$

(II) *Ber. dtsch. chem. Ges.* **44**, 2445 (1911):

The two compounds that can be represented by the diagrams

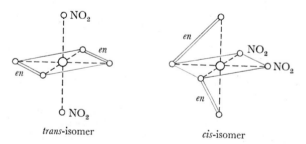

trans-isomer   *cis*-isomer

were prepared, and it was demonstrated that the *trans*-compound was optically inactive and non-resolvable, while the other readily gave optical isomers.

(IV) *Ber. dtsch. chem. Ges.* **44**, 3279 (1911):

In this paper it was proved that the complex $\begin{bmatrix} Cl \\ Cl \end{bmatrix} Co(en_2) \end{bmatrix}^+$, the analogue of the cis-form in II, could also be resolved:

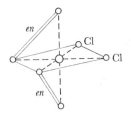

(V) *Ber. dtsch. chem. Ges.* **45**, 121 (1912)

In this final paper the dissymmetrical property of an octahedral complex was reduced to its simplest possible terms, in the complex $[Co(en)_3]^{3+}$,

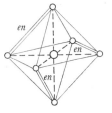

of which the optically active forms were prepared.

Concerning his fruitful application of the phenomenon of dissymmetry Werner wrote:*

Als wichtige Ergebnisse unserer bisherigen Untersuchung dürfen hervor gehoben werden:

1. der Nachweis, dass Metallatome als Zentralatome stabiler, asymmetrisch gebauter Moleküle wirken können;

2. der Nachweis, dass auch reine Molekülverbindungen in stabilen Spiegelbildisomeren auftreten können, wodurch der vielfach noch aufrecht erhaltene Unterschied zwischen Valenzverbindungen und Molekülverbindungen vollkommem verschwindet:

3. die Bestätigung einer der weitgehendsten Folgerungen aus der Oktaederformel, wodurch die letztere eine neue, wichtige Bestätigung gefunden hat.

(Attention may be called to the following as the principal results of these researches up to the present—

1. The proof that metal atoms can function as 'central atoms' of stable, dissymmetrical molecules;

2. The proof that wholly 'molecular compounds' can exist as stable, mirror-image isomers, by which the difference often asserted to exist between valence compounds and molecular compounds completely disappears.

3. The confirmation of the most far-reaching deductions from the octahedral model, which thus finds a weighty ratification.)

* *Ber. dtsch. chem. Ges.* **44**, 1887 (1911).

BIOGRAPHICAL AND OTHER REFERENCES

'Alfred Werner.' P. Karrer. *Helv. Chim. Acta*, **3**, 196 (1920).

Chemical Society Obituary: *J. chem. Soc.* **117**, 1639 (1920).

New Ideas in Inorganic Chemistry. A. Werner (trans. E. P. Hedley, with authorization and collaboration of the author, from the 2nd German edition (1908) of 'Neuere Anschauungen auf dem Gebiete der anorganische Chemie'). (London, 1911.)

'Beitrag zur Konstitution anorganische Verbindungen.' A. Werner. *Ostwalds Klassiker*, no. 212. (Leipzig, 1924.) (Annotations by P. Pfeiffer.)

8

VALENCY IN THE
TWENTIETH CENTURY

I have no hesitation in saying that, from a philosophical point of view, I do not believe in the actual existence of atoms, taking the word in its literal signification of indivisible particles of matter. I rather expect that we shall someday find, for what we now call atoms, a mathematico-mechanical explanation, which will render an account of atomic weight, of atomicity, and of numerous other properties of the so-called atoms.

> F. A. KEKULÉ, 'On some points of chemical philosophy:
> II. On the existence of chemical atoms'
> *The Laboratory,** **1**, 27 July 1867)

The theory of valency which was given by Heitler and London and by others has the great advantage of leading exactly to the concept of valency which is used by the chemists. But it seems questionable to me whether the quantum theory would have found or would have been able to derive the chemical results about valency, if they had not been known before.

> W. HEISENBERG, from an address to the Chemical Section of the
> British Association, 1931

Throughout the greater part of the last century physicists and chemists were strongly inclined to preserve the intellectual independence of their sciences: neither body was disposed to adopt ideas and speculations which had not been cultivated in its own field. Dalton, with prior conviction acquired from the writings of Newton, successfully convinced most chemists of the truth of the atomic theory of matter only because he devised, in the laws of constant and multiple proportions, a purely *chemical* proof. But, with some notable exceptions in Ampère and Clausius, the atomic theory and its consequences were little regarded by physicists. Faraday, who might be thought to have

---

* This journal, to which Kekulé contributed more than one article, in English, had a short life of less than a year (see also p. 2).

had a forceful incentive to incorporate the theory in his laws of electrolysis, expressed in 1853 the opinion

I do not know that I am unorthodox, as respects the atomic hypothesis. I believe in matter and its atoms as freely as most people, at least I think so. As to the little solid particles, which are supposed by some to exist independent of the forces of matter...as I cannot form any idea of them apart from the forces, so I neither admit nor deny them. They do not afford me the least help in my endeavour to form an idea of a particle of matter. On the contrary they greatly embarrass me....*

Such views, expressed by such a master of electrical experiment and discovery, just half a century after Dalton in 1803 had given his first table of atomic weights, offered no hope that at that time the notion of extending the atomic idea to *electricity* had even germinated. It was indeed nearly thirty years later that Helmholtz, delivering the Faraday Lecture to the Chemical Society 'On the Modern Development of Faraday's Conception of Electricity' in 1881, used the following words:

It was not felt necessary to accept any definite opinion about the ultimate nature of the agent which we call electricity. Faraday himself avoided as much as possible giving any affirmative assertion regarding this problem, although he did not conceal his disinclination to believe in the existence of two opposite electric fluids....Faraday's law of equivalence between quantity of electricity and extent of electrolytic decomposition tells us that through each section of an electrolytic conductor we have always equivalent electrical and chemical motion. The same definite quantity of either positive or negative electricity moves always with each univalent ion, or with every unit of affinity of a multivalent ion: this quantity we call the electric charge of the atom. Now the most startling result of Faraday's law is perhaps this. If we accept the hypothesis that elementary substances are composed of atoms, we cannot avoid concluding that electricity also, positive as well as negative, is divided into definite elementary portions, which behave like atoms of electricity.

[Of the discharge of hydrogen by electrolysis] the gas formed is electrically neutral: either every single atom is neutralized, or one atom remaining positive combines with another charged negatively. This latter assumption agrees with the inference from Avogadro's law that the molecule of free hydrogen is really composed of two atoms.

In his at that time daring proposals Helmholtz had in fact been forestalled ten years earlier by W. Weber, the founder of

* See A. J. Berry, *Modern Chemistry* (Cambridge, 1946), p. 30.

the theory of ferromagnetism, who in explanation of the source of this phenomenon postulated by Ampère in 1825, wrote in 1871:

The relation of the two particles (of positive and negative electricity) as regards their motions is determined by the ratio of the masses e and e' on the assumption that in e and e' are included the masses of the ponderable atoms which are attached to the electrical atoms. Let e be the positive electrical particle. Let the negative be exactly equal and opposite and therefore denoted by $-e$ (in place of e'), but let a ponderable atom be attracted to the latter so that its mass is thereby so greatly increased as to make the mass of the positive particle vanishingly small in comparison. The particle $-e$ may then be thought of as at rest and the particle $+e$ as in motion about the particle $-e$. The two unlike particles in the condition described constitute then an Amperian molecular current.

If the signs of the 'particles' be interchanged this representation becomes strikingly modern.

Maxwell, writing on 'Electrolysis' in *Electricity and Magnetism* in 1873, was still inclined towards Faraday's scepticism:

for convenience in description we may call this constant molecular charge revealed by Faraday's experiments one molecule of electricity,...it is extremely improbable that when we come to understand the true nature of electrolysis we shall retain in any form the theory of molecular charges.

This was an infelicitous prophecy, for the electron was discovered by learning the 'true nature' of gaseous electrolysis.

In the year in which Helmholtz delivered the Faraday Lecture Johnstone Stoney ventured to estimate the charge of an 'atom of electricity'. For the ratio F/N, where F is the faraday and N the number of atoms in one gram of hydrogen he obtained the value $e = 0.3 \times 10^{-10}$ e.s.u.* He was concerned to draw attention to the three fundamental units of physical measurement (1) the velocity of light, (2) the coefficient of gravitation, and (3) the elementary electric charge. About the last, to which he gave the name *electron*, he wrote

Nature presents us with a single definite quantity of electricity which is independent of the particular bodies acted on. To make this clear I shall express Faraday's law as follows: For each chemical bond which is ruptured within an electrolyte a certain quantity of electricity traverses the electrolyte, which is the same in all cases.

* *Phil. Mag.* (5), **11**, 384 (1881).

That the conception of the atomic nature of electricity had become accepted in mathematical physics before the actual discovery of the (negative) electron is shown in two papers by Larmor. In 'A dynamical theory of the electric and luminiferous medium', Part I, in the section, 'Introduction of Free Electrons'* he uses the term 'electron' in Stoney's sense of an atom of either positive or negative charge, and refers to one as 'the optical image' of the other. In a later communication 'On the theory of moving electrons and electric charges'† Larmor wrote

The facts of chemical physics point to electrification being distributed in an atomic manner, so that an atom of electricity, say an electron, has the same claims to separate and permanent existence as an atom of matter....Whatever view one may entertain as to the presence of qualities other than electric in an atom, all are, I think, nowadays agreed that the electron is there.

After a century of development in sometimes rather haughty isolation the sciences of physics and chemistry were forcibly drawn together again in consequence of the unparalleled series of experimental and theoretical advances following so swiftly upon each other during the twenty years 1894–1914. The chronology was as follows:

1894–1898	The discovery of the inert elements (Rayleigh and Ramsay)
1895	The recognition of 'X' radiation (Roentgen)
1895–1897	The isolation of the (negative) electron (Thomson)
1896	The discovery of radioactivity (Becquerel)
1898	The isolation and study of radium and polonium (M. and Mme Curie)
1900	The promulgation of the quantum theory (Planck)
1912	The diffraction of X-radiation by crystals (Laue)
1911–1913	The postulate of the nuclear atom (Rutherford and Bohr)
From 1913	The interpretation of line spectra upon the basis of atomic constitution (Bohr)

By 1916, through the work of Rutherford and Bohr, the concept of the 'nucleus atom' had been set upon a firm basis. Since the discovery that, under suitable conditions, free (negative) electrons could be expelled from all atoms, it was everywhere agreed that these particles must be discrete atomic constituents (cf. Larmor's views, above). But the electroneutrality of the atom was not secured by the presence of an equal number of

* *Phil. Trans.* **185**, 719–822 (1894). † *Phil. Mag.* **42**, 201 (1896).

'positive electrons' as imagined in the nineteenth century, nor by the sphere of uniform positive electrification in which Thomson had supposed the individual electrons to be embedded. Instead the whole positive charge (and almost all the mass) of the atom was borne by a single central particle or *nucleus*, of extreme minuteness when compared with the size of the whole atom. Thus all atoms contained electrons of identical mass and (negative) charge, but each atom carried its characteristic nucleus. The lightest atom, hydrogen, was presumed to contain, as its nucleus, the lightest known positively charged particle, and until recently this particle, named the *proton*, although highly 'unsymmetrical' to the electron, was the only recognized natural unit of positive electricity.

It was significant of the increasingly fertile consort of physics and chemistry that the period of steady development of theoretical chemistry should have been inaugurated by the publication in 1916, almost simultaneously and certainly independently, of two papers, the authors of which were a distinguished physicist and a chemist already of international renown: 'Über Molekülbildung als Frage des Atombaus' by W. Kossel* and 'The Atom and the Molecule' by G. N. Lewis.† These communications naturally bore a characteristic distinction of outlook: the one seeing (with Berzelius) all chemical union in terms of a generalized electrostatic attraction between electrically charged atoms; and the other, with a more practised intuition and more familiar knowledge of the large manifold of chemical problems, seeking in the well-tried chemical tradition to establish an empirical atomic model.

The following properties of the nuclear atom were soon established, and formed the bases of Kossel's and Lewis's expositions:

(1) The atomic number, Z, that is, the ordinal number of an element in the periodic system from $H = 1$ to $U = 92$, is equal

---

* *Ann. Phys. Lpz.*, **49**, 229, recd. Dec. 1915. Kossel was, from 1921, Professor of Theoretical Physics at the University of Kiel.

† *J. Amer. chem. Soc.* **38**, 762; recd. Jan. 1916. Lewis was Professor of Chemistry in the University of California.

to the charge, in protonic units, upon the nucleus of the atom; and therefore an atom of number Z contains Z electrons.*

(2) The atomic electrons are arranged in definite and successive groups or *shells*, distinguished by a discontinuous decrease of the strength of binding to the nucleus.†

(3) There exist at specific positions in the periodic system inert elements, which are incapable of chemical combination, and their electronic arrangements are presumed to have reached a maximum stability towards chemical influences.

The following is a synopsis of Kossel's paper.

Jedes folgende Element soll ein Elektron und ein Elementarquant positiver Ladung mehr enthalten als das vorhergehende. Zunächst zeigt allgemein die Tatsache des periodischen Wechsels des Valenzzahl dass beim Durchschreiten der Elemente von niederen zu höheren Gewichten die Konfiguration sich nicht gleichförmig ändert.... Vielmehr werden in regelmässigem Wechsel Konfiguration erreicht, in denen die Zahlen der zur Valenztätigkeit fähigen Elektronen sich wiederholen, u.a. auch solche, in denen praktische keine Niegung zum Austausch besteht, die Edelgase.... Wir fassen die Valenzeigenschaften ausdrücklich als eine Aussage über das Verhalten der äussersten Elektronen des Atoms auf.

(*Each successive element contains one electron and one elementary positive charge more than its predecessor. The fact that valency changes periodically proves at once that in passing from elements of lower to those of higher atomic weight the (electronic) configuration does not change uniformly. Instead, at regular stages, configurations are reached in which the number of electrons active in determining the valency is repeated: configurations associated with chemical inertness also recur regularly: these are those of the rare gases. We conceive the property of valency to be essentially an aspect of the behaviour of the outermost electrons of an atom.*)

Elements which are neighbours of the inert elements tend to gain or lose electrons until they attain the same number as the inert element. The forces of 'affinity' in the polar 'molecules' resulting are in fact electrical attractions between ions.

In the succession of elements from He to Ti elements of similar chemical properties and the same valency recur after *eight* places. The addition of eight electrons therefore leads to a similar ordering of electron 'rings' at the periphery of the atom: when a ring is full, as in the inert elements Ne and A, a new one is originated. In Na and K the ring is newly completed, and

* From the K lines of X-ray spectra: van der Broek, *Phys. Z.* **14**, 32 (1913); Moseley, *Phil. Mag.* **26**, 1024 (1913); *ibid.* **27**, 703 (1914).

† From the study of general X-ray spectra: Kossel, *Verh. dtsch. phys. ges.* **16**, 898, 953 (1914); *ibid.* **18**, 339 (1916).

the atoms of these metals contain one electron of a new ring, which is easily released: on the other hand, the atoms of F and Cl lack one electron for completion of their rings. As the position of an element recedes from or progresses beyond an inert element the tendency to gain or release electrons diminishes, but O and S may each take up two electrons, while N and P may each gain three. The alkali metals release one electron and the alkaline earth metals two.

Following the earlier suggestion of Abegg, Kossel uses the term *positive electrovalency* for the number of electrons easily released by an element, and *negative electrovalency* for the number absorbed. For any element from He to Ti the sum of the positive and negative electrovalency numbers must always amount to eight. Thus there exist NH_3, N_2O_5 : PH_3, PF_5 : ClH, Cl_2O_7. A simple explanation is thus provided for the rule first emphasized by Mendeleef (p. 85). In supposing that the positions adopted in a molecule by the valency electrons are decided exclusively by general electrostatic forces Kossel is compelled to ignore both the relation of valency to definite chemical bonds, and its directional properties. In fact in his view valency is reduced to a mere number (*Valenzzahl*) registering the electronic transactions of an atom.

Kossel was quite aware of the difficulties his theory would encounter in treating the constitution of symmetrical diatomic molecules. He accepts for H_2 the simple arrangement previously proposed by Bohr* 'Um den Mittelpunkt der Verbindungsachse der beiden Kerne kreisen die beide Elektronen in gemeinsamer Bahn', but rejects a similar model for HCl, which he treats as a heteropolar compound in which the total eight electrons are allocated to a ring in the plane of the chlorine nucleus.

Gehen wir nun schliesslich zu N_2 über, so ist klar, dass hier völlige Symmetrie eintreten wird. Gleichzeitig ist die Möglichkeit von einem positiven und einem negativen Partner zu sprechen, völlig verschwunden; entsprechend kann an elektrolytische Dissoziation nicht gedacht werden. Insgesamt hat das durch den Zusammenfluss der äusseren Ringe entstandene Gebilde hier zehn Elektronen zu enthalten....Noch mehr gilt dies für O_2 and Cl_2 in denen insgesamt 12 and 14 Elektronen zur Verfügung stehen....

* *Phil. Mag.* **26**, 1 (1913).

Gerade für diese Moleküle kann nun aber die Dispersionstheorie Aussagen machen. So hat Sommerfeld kürzlich (1915) gezeigt, dass das Verhältnis der Konstanten der Cauchy*schen* Dispersionsformel für N_2 und O_2 befriedigend von der Annahme wiedergegeben wird, der gemeinsame Ring enthalte bei N_2 sechs, bei O_2, vier Elektronen.

(*In the nitrogen molecule complete symmetry must prevail: the possibility of a negative and positive partner in the molecule has vanished, and electrolytic dissociation is unthinkable. Ten electrons are assembled by the confluence of the outer rings of each atom; for O_2 and Cl_2 the even larger numbers of 12 and 14 electrons are similarly to be considered. Sommerfeld's recent results from the application of the Cauchy dispersion formula to nitrogen and oxygen were in good agreement with a ring of six electrons held in common in N_2, and one of four electrons in O_2*).

Kossel, however, did not regard Sommerfeld's deductions with sufficient confidence to take the further step to the principle of shared electrons (or 'covalence'), which, first enunciated by Lewis, has remained a permanent tenet of valency theory.

Lewis opens his paper, on 'The Atom and the Molecule', by calling attention to the two classes, the *polar* (or ionizable) and the *non-polar* (or non-ionizable) into which chemical compounds can be broadly divided: but expresses the opinion that this classification is no more rigorous than others in natural science, in that partially ionizable bodies such as ammonia, water and acids connect the extreme non-polar compounds such as hexane with the fully ionized condition of a fused salt.

His theory of valency and chemical combination derives from four postulates:

(1) Every atom contains a (positively-charged) *kernel* which remains unaltered in all ordinary chemical actions. The positive charge, in protonic units, is equal to the number of the group of the periodic system in which the element lies.

(2) An atom is composed of the kernel and an outer shell which contains n electrons. In consequence of chemical change this number may vary from zero to a maximum of eight.

(3) The number n tends to be *even*, and especially an atom tends to the condition in which n reaches its maximum of eight. The n electrons in a shell are normally disposed at the corners of a *cube*, surrounding the kernel. Electrons can pass easily from one position to another in the shell.

(4) Two atomic shells are mutually interpenetrable.

Upon these postulates Lewis based his working model of the static, 'cubical' atom, which, in common with many previous models in chemistry, could clearly not be justified upon any actual physical grounds, but led nevertheless in Lewis' hands to qualitative concepts of valency which have survived the tests of modern physical and chemical advances.

The positive charge of atomic kernels, and the number of electrons in atomic shells

H, Li	Be, Mg			N, P			
Na, K	Ca, Sr	B, Al	C	As, Sb	O, S	Cl, Br	Ne, Ar
Rb, Cs	Ba	Sc	Si	Bi	Se, Te	I	Kr, Xe
1	2	3	4	5	6	7	8

Since Na (etc.) = (Ne (etc.) as kernel) $+ n$ electrons, the atoms of the inert elements are themselves kernels. In elements other than those listed the kernel is apparently neither uniquely determined, nor invariable during chemical action. In He two electrons in the shell play the same part as eight in the other inert elements, and in consequence we find hydrogen appearing in polar compounds as H^+ or as H^- (as in LiH). Hydrogen is thus analogous to both the alkali metals and the halogens.

In relation to postulate (3) the molecules NO (11), NO_2 (17), and ClO_2 (19) seem to comprise the only examples of the rare case of an *odd* total number of electrons in the shells of combined atoms.

Postulate (4) yields the clue to the constitution of non-polar compounds, which clearly cannot be explained by a complete transference of electrons from one atom to another:

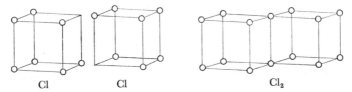

Cl   Cl Cl_2

In the formation of the chlorine molecule so expressed the shell of each atom attains its maximum content of eight electrons. As a convenient symbol for the pair of electrons shared in such a molecule the colon sign placed between the atomic symbols is suggested:

Cl:Cl.

It is evident that the type of union which we have so pictured, although it involves two electrons held in common by two atoms, nevertheless corresponds to a single bond as it is commonly used in graphical formulae. In order to illustrate this point further we may discuss a problem which has proved extremely embarrassing to a number of theories of valence: I refer to the structure of ammonia and the ammonium ion. This ion may, of course, on account of the extremely polar character of ammonia and hydrogen ion be regarded as a loose complex due to electric attraction, but as we consider the effect of substituting hydrogen by organic groups we pass gradually into a field where we may be perfectly certain that four groups are attached directly to the nitrogen atom, and held with sufficient firmness so that stereo-chemical isomers have been obtained. We explain the combination of ammonia with hydrogen ion by the equation

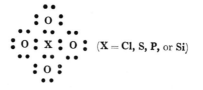

in which the bold-type symbols denote atomic kernels. The total positive charge upon the kernels is $5+4 = 9$, while the shells contain a total of eight electrons: the ammonium ion has thus necessarily one positive charge. In ammonium chloride the chlorine ion is not attached directly to nitrogen but is held simply by electrostatic force.

While the two dots of our formulae correspond to the line which has been used to represent a single bond, we are led through their use to certain formulae of great significance which I presume would not occur to anyone using the ordinary symbols....We may now write formulae in which an atom of oxygen is tied by only one pair of electrons to another atom and yet have every element in the compound completely saturated. For ClO_4^-, SO_4^{2-}, PO_4^{3-} and SiO_4^{4-} we can write uniformly*

$$\overset{\bullet\bullet}{\underset{\bullet\bullet}{\colon O \colon}}$$

$$\colon O \colon X \colon O \colon \quad (X = Cl, S, P, \text{ or } Si)$$

$$\colon O \colon$$

The formation of sulphate ion by the union of sulphur trioxide with the oxygen ion is completely analogous to the formation of ammonium ion

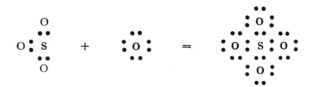

* Atomic kernels are indicated by bold face symbols.

Lewis recognized two shortcomings of his cubical model: it does not provide for the essential chemical requirement of free rotation about a single bond; and while it is compatible with the formation of double bonds, as in the oxygen molecule

O_2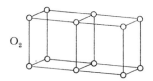

it cannot accommodate triple bonds. Noting again that in He a pair of electrons replaces the octet of electrons in succeeding inert elements, he inclines to the view that the stable electron pair, as in helium, may be a more fundamental unit than the octet group. For carbon (and other small atoms) in chemical saturated states he therefore proposes a model

with only tetrahedral symmetry, which allows the formation of triple bonds and also free rotation round a single bond.

Lewis' lasting contribution to the explanation of valency as exhibited in non-polar (or *covalent*)* compounds was his concept of shared electron pairs, and not the static model by means of which he found it convenient to illustrate that concept. 'In developing the theory of valency the chemist may use symbols with no definite physical connotation to express the reactivity of the atoms in a molecule, and may leave it to the subsequent progress of science to discover what realities these symbols represent.'† Today the prevalence and mode of operation of

---

* The convenient term *covalent*, to describe a chemical link brought into being by the sharing of electrons supplied either by one or both of the linked atoms, appears to have been first suggested by I. Langmuir. The term soon became used also to describe a compound owing its existence to such interatomic links.

† N. V. Sidgwick, from the Preface to *The Electronic Theory of Valency* (Oxford, 1927).

electron sharing has developed into a principal part of theoretical chemistry.

It was therefore some disservice to the development of a rational and physically well-grounded theory of valency that many chemists, under the lead especially of I. Langmuir,* should seize upon the wholly pictorial and inevitably ephemeral part of Lewis' thesis, the 'cubical atom' and its attendant 'octet theory of valency', to devise a range of fanciful chemical structures largely irreconcilable with any known physical principles, and in which 'atoms lay around like dry-goods boxes in a packing-shed' if one may be allowed to recall the satirical comment of a famous physicist of the time. The octet theory, in which it was assumed that atoms *in general* became chemically saturated when by any means eight electrons had been acquired in their valency 'shells', incited a temporary resuscitation of the doctrine of constant valency, of which Kekulé had lived to be the sole advocate. For example, the molecular constitution of the oxychlorides of sulphur could be built up from that of sulphur dichloride, SCl_2, in which the element was clearly bivalent and had already acquired an 'octet', by the successive addition of two atoms of oxygen each sharing two electrons from the sulphur octet

$$S^{2-} \qquad SCl_2 \qquad SOCl_2 \qquad SO_2Cl_2$$

Throughout this series the atom of sulphur appeared to maintain the same 'saturated' electronic environment, and therefore the same bivalency. By similar means phosphorus could be accorded a uniform tervalency in the series PCl_3, $POCl_3$ and PO_4^{3-}.

In a second paper, entitled 'Isomorphism, Isosterism, and Covalency'† Langmuir enunciated a principle resting upon

* *J. Amer. chem. Soc.* **41**, 868 (1919).
† *J. Amer. chem. Soc.* **41**, 1543 (1919).

more plausible physical foundations: 'if compounds having the same number of atoms have also the same total number of (valency) electrons, the latter tend to arrange themselves in the same manner'. Compounds related in this way he termed 'isosteric'. Mitscherlich, who originated the principle of crystallographic isomorphism, had assumed that it concerned only 'compounds of similar chemical formulae composed of elements of similar chemical properties', but it had come to be recognized that many strictly isomorphous pairs, such as calcite, $CaCO_3$ and sodium nitrate, $NaNO_3$; or barium sulphate, $BaSO_4$ and potassium perchlorate, $KClO_4$, imply some more comprehensive principle. Such an extension is offered by Langmuir's assumption: the valency shells of calcium and carbon together contain $2+4$ electrons, while those of sodium and nitrogen hold $1+5$ electrons. Similarly potassium with chlorine $(1+7)$ and barium with sulphur $(2+6)$ supply the sum of eight electrons.

Langmuir successfully applied the principle of 'isosterism' to correct the structural formulae of molecules and radicals. The oxides carbon dioxide, CO_2, nitrous oxide, N_2O, with the azide radical —N_3 and radicals of the general formula (CNO)— all comprise a total of sixteen valency electrons. Carbon dioxide, according to the established principles of the stereochemistry of carbon, must possess a structural formula $O{=}C{=}O$ in which the centres of the three atoms are collinear. It should follow that the formula of nitrous oxide is not N⚊⚊⚊N nor that of

azide and the metrical methods of the present time have amply confirmed the formulae $N{=}N{=}O$ and $(N{=}N{=}N)$— originally predicted by Langmuir's principle. The cyanate, isocyanate and fulminate radicals receive similar linear arrangements $(N{=}C{=}O)$—, —$(N{=}C{=}O)$, and —$(O{=}N{=}C)$ respectively, which have also been confirmed by present-day methods.

It is to the keen perception of N. V. Sidgwick* that we owe

* *J. chem. Soc.* **123**, 725 (1923).

the successful application of Lewis' principles to the large class of 'coordination compounds' so fully explored and firmly established by Werner. In their normal fluorides, BeF_2 and BF_3, the atoms of beryllium and boron are unable to reach electronic saturation in their valency shells, since after these compounds have once been formed with elementary fluorine neither atom possesses further electrons to contribute to shared electron pairs. Saturation is, however, attained by action with fluoride *ions*, F^-, which themselves supply the electron pairs required:

$$F \overset{\bullet}{\underset{\bullet}{}} Be \overset{\bullet\bullet}{\underset{\bullet\bullet}{F}} \quad + \quad 2(\overset{\bullet}{\underset{\bullet}{}} F) \quad = \quad F \overset{\bullet}{\underset{\bullet}{}} Be \overset{\overset{\textstyle F}{\bullet\bullet}}{\underset{\bullet\bullet}{\underset{F}{}}} \overset{\bullet}{\underset{\bullet}{}} F$$

$$\text{or,} \quad BeF_2 \quad + \quad 2F^- \quad = \quad BeF_4{}^{2-}$$

and similarly with boron trifluoride:

$$BF_3 \quad + \quad \overset{\bullet}{\underset{\bullet}{}} F \quad = \quad BF_4{}^-$$

Lewis had already discussed the rationale of the formation of ammonium ion, which Werner had looked upon as one of the simplest coordination compounds.

No anions of beryllium or boron containing more than four fluorine atoms are known, and we therefore conclude that in the series

$$BeF_4{}^{2-}, \; BF_4{}^-, \; CH_4 \text{ or } CF_4, \; NH_4{}^+$$

the valency shells are saturated with an octet of electrons, implying a uniform quadri-covalency, or in Werner's terminology a co-ordination number of 4. If we turn to the next Series of the Periodic System, aluminium fluoride, AlF_3, is unsaturated similarly to BF_3, but if we retain the octet principle we must regard silicon tetrafluoride, SiF_4, as saturated, like CF_4, and incapable of further addition. Nevertheless, silicon tetrafluoride readily takes up two fluoride ions to yield the stable ion $SiF_6{}^{2-}$, while aluminium trifluoride combines with as many as three fluoride ions to form the very stable anion $AlF_6{}^{3-}$. Each of these ions displays the coordination number 6, which Werner had found to be so prevalent.

These examples led logically to the general assumption that, as well as the halogen ions, molecules or ions, such as ammonia, NH_3, water, H_2O, or nitrite ion, NO_2^-, in which one or more pairs of electrons remain unshared, could function as 'ligands' in a co-ordination complex. The otherwise puzzling observation of Werner that an uncharged molecule such as ammonia can be replaced in metallic complexes by numerous ions unit for unit is thus easily explained. In entering a complex an ion confers upon it its own charge and there results the characteristic series, with constant co-ordination number but steadily changing ionic charge. The complexes formed by platinic chloride and ammonia with empirical formulae $PtCl_4 . nNH_3$ and co-ordination number 6 provide an example of such series in which all the members have been prepared:

$$[Pt(NH_3)_6]^{4+} . Cl^-_4 \qquad [Pt(NH_3)_5Cl]^{3+} . Cl^-_3,$$

$$[Pt(NH_3)_4Cl_2]^{2+} . Cl^-_2 \qquad [Pt(NH_3)_3Cl_3]^+ . Cl^-,$$

$$[Pt(NH_3)_2Cl_4]^0$$

$$[Pt(NH_3)Cl_5]^- . K^+ \qquad [PtCl_6]^{2-} . K^+_2.$$

Sidgwick commemorated in the term 'co-ordinate link' the success of his discernment in co-ordination compounds of a link in which a 'donor' atom A supplies both electrons to attach an 'acceptor' atom B, and he originated the use of the sign A→B, in place of A—B, to denote the unsymmetry of the electron supply. Since there is no reason to suppose that a co-ordinate link once formed would be less stable than any other covalent link, it is to be expected that a co-ordination complex would, like the compounds of carbon, be subjected to rules of stereochemistry, as Werner so brilliantly demonstrated by preparing numerous complexes as optical isomerides.

It is clear that there is no real opposition between the Werner theory and that of structural chemistry...and the reason why the structural theory, in spite of disregarding them [co-ordination compounds] has been of so much service in organic chemistry is very largely that suggested by Werner, that carbon, having the number of valency electrons required to give it directly its maximum covalency of four, seldom if ever forms coordinate links.... This interpretation also removes an apparent weakness in the Werner theory, that it seemed to attribute a unique character to one atom in the molecule, as the 'centre of coordination'. It is now evident that this is merely an arbitrary distinction, a matter of convenience.... Every atom has a valency group of electrons made up by the same processes to a stable number.*

In 1923 the Faraday Society organized 'A General Discussion on the Electronic Theory of Valency', which was held at Cambridge under the chairmanship of Sir J. J. Thomson.† G. N. Lewis introduced the proceedings with a paper entitled 'Valency and the Electron'. He recalled that Bohr in the preceding year had accounted for the general characteristics of the Periodic System by assuming groups of energy levels: the first may contain 2 electrons (He), the second 8, the third 18, and the fourth 36, in agreement with the formula

$$Z = 2(1^2 + 2^2 + 2^2 + 3^2 + 3^2 + 4^2 + \ )$$

given by Rydberg in 1914 for the atomic numbers of the inert elements. His theory however gave no clue to the importance of groups of 2 or 8 electrons in the outermost shells.

I called attention some years ago to the remarkable fact that when we add up the valency electrons...all but a handful of some 100,000 known substances contain an *even* number of electrons. The universality of this pairing of electrons points definitely to an actual physical coupling of electron orbits. Every atom or molecule with an odd number of electrons possesses a magnetic moment, while the great majority with an even number have no such moment.

It would be going a little too far to say that it is magnetic force which couples two electronic orbits to form the electron pair. For the present it will be sufficient to assert that the coupling of two electron orbits with neutralization of their magnetic fields is the most fundamental of chemical phenomena.

Lewis referred here to the magnetic fields associated with the *orbital* rotations of electrons: the concept of electron *spin* was not recognized until 1925.

In a short communication under the title 'The Nature of the Non-polar Link' N. V. Sidgwick remarked 'The only new

* Sidgwick, *Electronic Theory of Valency*, p. 115.
† *Trans. Faraday Soc.* **19**, 450.

assumption I make is that the orbit of each shared electron includes both of the attached nuclei'. He conceived that a link of such a nature would have axial symmetry and would therefore allow the necessary free rotation of single links. Moreover, the need of *two* electrons in forming a link might be explained, for while the one is in a 'perihelion' position and exerting little or no binding force between the atoms the other might occupy a central position where its maximum binding force would be exerted. In the discussion another member of the conference made the interesting suggestion that the binding orbits might take the form of a figure of eight.

In opening the second section of the discussion, devoted to the electron theory in organic chemistry, the President of the Faraday Society, Professor Robert Robinson,* said 'At present it would seem as if we are in a transition stage of knowledge comparable with that which obtained in the forties and fifties of last century, and the chemists of two or three generations hence will look back upon the present confusion with the same feelings as we experience in regarding that time'. 'The present confusion' so candidly acknowledged by the President arose of course because a sound physical basis for chemical speculations upon an electronic theory of valency was still lacking. The difficult task of elucidating atomic constitution under the guidance of the facts of spectroscopy had indeed made rapid progress, but the outcome still rested in the last resort upon an entirely empirical foundation, in conflict with what appeared to be well-established conceptions of physics.

In his original treatment of the hydrogen atom† Bohr had assumed three postulates:

(1) The electron, regarded as a negatively charged particle, rotated in circular orbits about the nucleus under a Coulomb field of force, i.e. with potential energy $V = -e^2/r$.

(2) The possible stable (and therefore non-radiating) orbits were given by the condition

$$p_\theta = mr^2\omega = nh/2\pi \quad (n \text{ a whole number}),$$

* Later Sir Robert Robinson, O.M.   † *Phil. Mag.* **26**, 1, 476 (1913).

where p_θ = angular momentum in the orbit, r = orbital radius, ω = angular velocity, n = quantum number. This postulate was derived from Planck's theory of the harmonic oscillator (1900).

(3) Spectral frequencies ν were given by Planck's condition

$$\nu = (E_2 - E_1)/h,$$

applied to transitions between stationary states of energy respectively E_1 and E_2.

From (2) the electronic energy was calculated as

$$E_n = \frac{2\pi^2 m e^4}{h^2} \frac{1}{n^2},$$

and consequently from (2) and (3) Rydberg's constant for hydrogen spectra was $R = 1 \cdot 09 \times 10^5$ cm.$^{-1}$. The almost perfect agreement with the accepted experimental value 109,675 cm.$^{-1}$ could not fail to lend the strongest support to the validity of the postulates. Nevertheless, the very existence of discrete non-radiating states of the electron remained physically incomprehensible.

After Sommerfeld had extended Bohr's formulation to include elliptical orbits,* which require two quantum numbers, n and k, it became possible to include atoms of higher atomic number, and in 1921 Bohr put forward a provisional electronic constitution for the inert gases:†

	Z	1	2		3			4				n
		1	1	2	1	2	3	1	2	3	4	k
He	2	2										
Ne	10	2	4	4								
A	18	2	4	4	4	4						
Kr	36	2	4	4	6	6	6	4	4			

* *Ann. Phys., Lpz.*, **51**, 1 (1916).
† *Aufsätze über Spektren und Atombau*, vol. III (Braunschweig, 1922).

It will be seen that the electrons associated with a principal quantum number n are arbitrarily assumed to be equally distributed over the subsidiary numbers k. The major axes of orbits are proportional to n^2 and their minor axes to nk: hence when $n = k$ the orbits are circular as in Bohr's original hydrogen model.

Bohr's scheme for the allocation of electrons in the inert elements depended upon two quantum numbers, n and k, the first of which takes the successive integral values 1, 2, 3, etc., while k takes the integral values 1, 2, ..., n. To each of the n subgroups of electrons within each 'shell' of principal number n he arbitrarily assigned an equal number of electrons. Consequently, as n increased, subgroups of the same k value contained an increasing number of electrons: for example for $n = 1, 2, 3$, the electrons assigned to the $k = 1$ level were respectively 2, 4, and 6. The electron content of a sublevel was thus made to depend upon both n and k.

E. C. Stoner* proposed the concept of 'closed' subgroups: in every shell ($n = 1, 2, 3$, etc.) the maximum number of electrons possible in each of the sublevels is fixed and cannot increase further with increase of n. This maximum number therefore depends only on k. Stoner gave the number $2(2k-1)$ for this maximum content

k	1	2	3	4
Electrons in closed groups	2	6	10	14

To explain the fine structure, i.e. doublets, etc., especially in the spectra of the alkali metals, he proposed a third quantum number m, which took only two values, $k-1$, and k. By this means each closed subgroup is divided into two sections, each composed of $2m$ electrons.

In 1925, in a paper entitled 'Ueber den Zusammenhang des Abschluss der Elektronengruppen im Atom mit der Komplexstruktur der Spektren',† the Swiss physicist W. Pauli enunciated in its first restricted form a principle which in its ultimate development ranks like the atomic theory as one of the fundamental bases of the organization of nature.

* *Phil. Mag.* **48**, 719 (1924).   † *Z. Phys.* **31**, 765 (1925).

Electronic distribution in the inert elements (Stoner)

							Number of n, k, m electrons								
Z	1_1	2_1	$2_{2(1+2)}$	3_1	$3_{2(1+2)}$	$3_{3(2+3)}$	4_1	$4_{2(1+2)}$	$4_{3(2+3)}$	$4_{4(3+4)}$	5_1	$5_{2(1+2)}$	$5_{3(2+3)}$	6_1	$6_{2(1+2)}$
He	2														
Ne	2	2	2+4												
A	2	2	2+4	2	2+4										
Kr	2	2	2+4	2	2+4	4+6	2	2+4							
Xe	2	2	2+4	2	2+4	4+6	2	2+4	4+6	—	2	2+4			
Rn	2	2	2+4	2	2+4	4+6	2	2+4	4+6	6+8	2	2+4	4+6	2	2+4
Present day (see p. 151)	1s	2s	2p	3s	3p	3d	4s	4p	4d	4f	5s	5p	5d	6s	6p

Note: the Z values for He, Ne, A, Kr, Xe, Rn are 2, 10, 18, 36, 54, 86 respectively.

144

In Stoner's scheme, account had not been taken of the striking development of fine structure in spectra of atoms placed in strong magnetic fields (the Paschen–Back effect). An electron in a given orbital state represented by the quantum numbers n, k, and m possesses a definite angular momentum and may therefore react to an imposed magnetic field in two ways depending on the two possible directions of the orbital rotation. Pauli therefore proposed to include this aspect of the electronic state in the form of a fourth quantum number m_1. The quantitative study of the effects of magnetic fields had shown that m_1 must be given half-integral values:

m	m_1	No. of magnetic states
1	$\pm 1/2$	2
2	$\pm 1/2, \pm 3/2$	4
3	$\pm 1/2, \pm 3/2, \pm 5/2$	6

After this addition Pauli could assign to every single electron in Stoner's scheme its own distinctive set of quantum numbers, and gave the following 'exclusion principle' governing the whole assignment of electronic states in an atom:

Es kann niemals zwei oder mehrere äquivalent Elektronen im Atom geben, für welche in starken Feldern die Werte allen Quantenzahl n, k, m, and m_1 übereinstimmen. Ist ein Elektron im Atom vorhanden, für das diese Quantenzahlen (in äusseren Felde) bestimmt Werte haben, si ist diese Zustand 'besetzt'.

(*There can never exist in an atom two or more equivalent electrons for which (in strong external fields) all the quantum numbers n, k, m and m_1 are the same. An atomic state containing an electron with all four numbers defined is completely occupied.*)

It can hardly be doubted that the assignment of all electrons in the inert elements to closed groups is the physical aspect of their chemical indifference. The efforts of Stoner and Pauli thus afforded, perhaps rather belatedly, support of the ideas brought forward by Kossel and Lewis some eight years previously.

Less than a year after the publication of Pauli's paper the Dutch spectroscopists Uhlenbeck and Goudsmit* revived an

---

* *Naturwissenschaften*, **13** (20 Nov. 1925); transcript *Nature, Lond.* **117**, 264 (1926).

idea upon which there had been previous speculation, namely, that an atomic electron possessed not only angular momentum due to its orbital motion but also an intrinsic momentum originating in spin round its own axis. In applying this concept to an explanation of the doublets in the alkali-metal spectra they found it necessary to attribute angular momentum $\pm\frac{1}{2}h/2\pi$ to this spin motion, in its two possible directions; whereby an electron state which on Bohr's theory was associated with n units $(h/2\pi)$ of orbital angular momentum split into two states with momenta $n \pm \frac{1}{2}$. In this application Uhlenbeck and Goudsmit found no grounds for attributing to the spin angular momentum a higher value than $\frac{1}{2}h/2\pi$. It appeared therefore that its spin was as constant a property of an electron as its mass or charge. The hypothesis of intrinsic electron spin provided a simple and natural explanation of the half-integral quantum numbers previously introduced for total electronic angular momentum upon purely empirical grounds.

While it was obvious that the hypothesis could have an important bearing upon the operation of Pauli's exclusion principle, and therefore upon the assignment of electron states in atoms, its full implication could not be developed until a physical basis of Bohr's theory had been revealed.

The theoretical advance which clarified and physically authenticated the whole outlook upon atomic constitution, and in due course also that of molecules, and thus ensured the supremacy of wave mechanics in attacking such problems, originated in two papers by L. de Broglie: 'A Tentative Theory of Light Quanta',* and 'Recherches sur la théorie des quanta.'† As antecedents to de Broglie's conceptions it must be recalled that Planck in initiating the quantum theory (1900) confined quantization to the interactions between radiation and matter, and predicated nothing about the nature of radiation itself. Five years later, Einstein pointed out that the phenomena of the photo-electric effect were incomprehensible unless it was assumed that radiation itself was quantized as 'corpuscles' of energy $h\nu$, to which he gave the name of *photons*. In opening his

---

* *Phil. Mag.* **47**, 446 (1924). † *Ann. Phys., Paris*, **3**, 65 (1925).

English paper de Broglie stresses the reality of light quanta, and recalls the recently discovered Compton effect (1922). In this phenomenon a high-energy photon (of X-radiation) collides, so to speak on equal terms, with an electron, and both 'particles' deviate from the collision at angles with the direction of the radiation determined by the classical principle of conservation of energy and momentum. Since the electron gains energy at the expense of the photon, the frequency associated with the latter is diminished, as experiment proved. In this effect therefore the photon behaves as a particle in a collision governed by classical laws, but after as well as before the event continues as radiation with definite though lessened, frequency, and increased wavelength.

This duality of aspect led de Broglie to write in the section of his English paper, entitled 'An Important Theorem of the Motion of Bodies'.*

We are then inclined to admit that any moving body may be accompanied by a wave, and that it is impossible to disjoin motion of body and propagation of wave....The theory suggests an interesting explanation of Bohr's stability conditions [for electron orbits]. At time o the electron is at a point A of its trajectory. The phase wave starting at this instant from A will describe all the path and meet the electron again at A'. It seems quite necessary that the phase wave shall find the electron in phase with itself. That is to say, the motion can only be stable if the phase wave is tuned with the length of the path, or

$$\int \frac{dS}{\lambda} = n \quad \text{(a whole number)}.$$

Since for a photon of energy $h\nu$ the momentum is $h\nu/c = h/\lambda$ (c the velocity of light), for 'matter' waves the relation $p = h/\lambda$ may be expected to hold generally. In his second paper de Broglie derives Bohr's second postulate (p. 141) from this relation:

putting $\lambda = 2\pi r/n$,

we have at once

$$2\pi r \cdot mv = 2\pi r \cdot m\omega r = nh,$$

and he expresses the opinion 'Ce beau résultat dont la démonstration est si immédiate...est la meilleure justification que nous puissions donner de notre manière d'attaquer le problème des quanta'.

* Phil. Mag. **47**, 450 (1924).

Direct experimental confirmation was forthcoming in 1927, when Davisson and Germer,* acting upon the suggestion of Elsasser in 1925,† demonstrated the diffraction of a beam of electrons reflected by a crystal of nickel in exact analogy with X-ray diffraction: independently, in the same year G. P. Thomson and A. Reid‡ proved a similar diffraction on transmission through thin films of metals, as in Laue's experiments with X-radiation. Particular care was taken to confirm de Broglie's relation

$$p = h/\lambda;$$

$p = mv$ is known from the potential accelerating the electron beam, and the associated wavelength by applying Bragg's law to the diffracted beam. In the year before the real existence of matter waves had so dramatically passed the test of experiment, Schrödinger had shown how de Broglie's equation, which in its simple form applies only to the freely moving electron, could be plausibly adapted to the case of electrons bound within atoms.§

The classical equation describing any wave propagation in the direction x is

$$\frac{d^2\psi}{dx^2} + \frac{4\pi^2}{\lambda^2}\psi = 0,$$

where ψ = amplitude of the wave; λ = the wave-length. For the wave associated with an electron travelling in the direction x

$$\lambda^2 = \frac{h^2}{p_x^2} = \frac{h^2}{2mT},$$

where p_x = momentum, T = kinetic energy, m = mass, of the electron. If the electron is bound to the nucleus of an atom we must have

$$T = E - V,$$

where E is the (constant) total energy of the electron-nucleus system, and V is the potential energy of the binding.

Expanding to three degrees of freedom in the motion, and inserting the calculated value of λ^2 we find

$$\nabla^2\psi + \frac{8\pi^2 m}{h^2}(E-V)\psi = 0, \quad \left(\nabla^2 = \frac{\partial^2}{\partial x^2} + \frac{\partial^2}{\partial y^2} + \frac{\partial^2}{\partial z^2}\right).$$

---

* *Phys. Rev.* **30**, 707 (1927). † *Naturwissenschaften*, **13**, 711 (1925).
‡ *Nature, Lond.*, **119**, 890 (1927). § *Ann. Phys., Lpz.*, **79**, 36 (1926).

Re-arranging, $\quad [V - (h^2/8\pi^2 m)\nabla^2]\psi = E\psi.$

This now generalized wave-equation may be concisely expressed as

$$H\psi = E\psi$$

where H is the operator enclosed in brackets.

The physical interpretation of ψ, a function of the co-ordinates x, y, z, and essentially the amplitude of electron waves, was first given by M. Born.* It may be deduced by analogy with the admitted duality of light. Regarded as a wave propagation the light intensity at any point is proportional to the square of the amplitude at the point: as a propagation of photons the point intensity is proportional to the point density of photons. Correspondingly ψ^2 is proportional to the electron density at a point in a freely moving beam of electrons. When the electron beam is reduced to the ultimate limit of one moving electron then ψ^2 at a point (x, y, z) must give the *probability* of finding the electron at the point. Now from the form of the wave-equation it is seen that if ψ is one solution, then $A\psi$ where A is any arbitrary constant is also a solution, but if the suggested interpretation of ψ^2 is adopted then necessarily

$$A^2 \int \psi^2 dv = 1$$

when the integration is over all co-ordinate space. Hence a function ψ is *normalized* on multiplication with the factor $1/A$ given by $\int \psi^2 dv = 1/A^2$.

In the hydrogen atom $V = -e^2/r$, where r is the radius vector of the electron about the proton. With this value of V and *any* arbitrary value of E, the total energy, at least one solution for ψ can be found which satisfies the wave equation. However, only for a few discrete energy values, *which coincide with those of Bohr's stationary states*, do the solutions present the following necessary properties: ψ must be everywhere finite (or zero); both ψ and its derivatives

$$\frac{d\psi}{dx}, \frac{d\psi}{dy} \quad \text{and} \quad \frac{d\psi}{dz}$$

must be everywhere continuous and single valued. Thus from the wave-equation alone, without any of the special hypotheses

* *Z. Phys.* **38**, 803 (1926).

that Bohr was compelled to construct, the correct energy states emerge. This and other similar predictions in agreement with experiment form the real basis of a 'proof' of the authenticity of the equation, about the derivation of which the sceptical might be disposed to entertain some qualms.

The number of independent and acceptable wave-functions appearing as solutions of the wave equation for hydrogen, and corresponding to the allowed energies Rch/n^2 (R = Rydberg's spectral constant) is equal to n^2:

Energy	No. of functions
E_1 (lowest, or ground state)	1
E_2	4
E_3	9
etc.	etc.

The number of orbital functions thus increases much faster than in the scheme of orbits proposed by Bohr and Sommerfeld, under which for each level n there exists one circular orbit and $n-1$ elliptical orbits, defined by the two quantum numbers n and k (p. 142). From the treatment by wave-mechanics it emerges that each function ψ is characterized by three quantum numbers, n, l, m:

n	l	m
1, 2, 3, etc., total energy, Rch/n^2	0, 1, 2, ... $(n-1)$, total angular momentum (units $h/2\pi$) $\sqrt{l(l+1)}$	$\pm l$, $\pm(l-1)$, ..., 0, magnetic moment in magnetic field (in magnetons)

The functions may first be classified by using only the numbers n and l.

l	Type of function	Occurrence
0	s	At all levels ($1s$, $2s$, etc.)
1	p	At levels $n > 1$ ($2p$, $3p$, etc.)
2	d	... $n > 2$ ($3d$, $4d$, etc.)
3	f	... $n > 3$ ($4f$, $5f$, etc.)

The convenient literal symbols s, p, d, etc., were drawn from long-established spectroscopic practice. Functions within these main classes s, p, etc., are then distinguished by the 'magnetic' quantum number m:

Type	m	No. of functions (electronic states)	No. including spin
s	0	1	2
p	± 1, 0	3	6
d	± 2, ± 1, 0	5	10
f	± 3, ± 2, ± 1, 0	7	14

The additional source of angular momentum and magnetic moment attributable to electron spin is extraneous to the functions predicted by the Schrödinger equation in its ordinary form, but is easily included, as a *fourth* quantum number, m_s, which can take only the two values $\pm \frac{1}{2}$.

		Stoner–Pauli		Schrödinger		
	n, k, m	m_1	Maximum electrons	$m_s = \pm \frac{1}{2}$ (spin)	m	n_l
Ne	1_1	$\pm \frac{1}{2}$	2	2	$\pm \frac{1}{2}$	0	$1s$
	2_1	$\pm \frac{1}{2}$	2	2	$\pm \frac{1}{2}$	0	$2s$
	$2_{2,1}$	$\pm \frac{1}{2}$	2	2	$\pm \frac{1}{2}$	0	$2p$
	$2_{2,2}$	$\pm \frac{1}{2}$, $\pm \frac{3}{2}$	4	4	$\pm \frac{1}{2}$, $\pm \frac{3}{2}$	± 1	
Kr	$3_{3,2}$	$\pm \frac{1}{2}$, $\pm \frac{3}{2}$	4	4	$\pm \frac{1}{2}$, $\pm \frac{3}{2}$	± 1	$3d$
	$3_{3,3}$	$\pm \frac{1}{2}$, $\pm \frac{3}{2}$ $\pm \frac{5}{2}$	6	6	$\pm \frac{1}{2}$ $\pm \frac{3}{2}$, $\pm \frac{5}{2}$	0 ± 2	

In the new theories therefore the essential number of features (or quantum numbers) distinguishing one orbital function from another in the hydrogen atom is identical with that proposed by Pauli for a more general use (p. 145). The Stoner–Pauli scheme (p. 144) had, of course, already included electron spin, in the form of empirical half-integral quantum numbers, without recognizing it as a separate part of the total electronic angular momentum. We now compare the distribution of electrons among the orbits in the Stoner–Pauli scheme for the inert elements with the distribution obtained if we assumed that the new classification determined for hydrogen could also be applied to these elements of higher atomic number.

The complete agreement in the possible electronic states, and therefore in the number of electrons assignable, can only mean that the scheme of classification arising from applying the wave-equation to hydrogen, when combined with the conception of electron spin, is also valid for the closed sets of electrons in the Stoner–Pauli scheme for the inert elements; and this latter, it is to be recalled, was founded directly upon experimental (spectroscopic) data. Two consequences follow: an increased confidence in the authenticity of wave-mechanical principles; a means of ascertaining the electronic structure of any atom of number Z, by what Bohr called the 'aufbau-prinzip'. The Z electrons are to be distributed, under Pauli's exclusion principle, over the number of states brought to light in the treatment of the hydrogen atom, with the condition that every electron is bound as strongly as possible.

The types of spectra emitted by the elements between and including hydrogen and krypton, and their spectroscopic ground-states so deduced,* enable Series I–IV of the Periodic System to be empirically subdivided as in columns 1–3 below:

	No. of elements	States of added electrons
I H, He	2	$1s$
II { Li, Be	2	$2s$
{ B–Ne	6	$2p$
III { Na, Mg	2	$3s$
{ Al–Ar	6	$3p$
IV { K, Ca	2	$4s$
{ Sc–Cu	10	$3d$
{ Zn–Kr	6	$4p$

* See Palmer, *Valency, Classical and Modern*, ch. v.

The significant numerical succession in column 3 can hardly leave any doubt that column 4 shows the order of occupation of atomic orbitals under the *aufbau-prinzip*, and therefore the order of strongest binding (in the atomic ground-states).

In hydrogen all orbitals of the same principal quantum number n are *degenerate*, that is to say, their associated energies of electron binding are identical. This maximum possible degree of degeneracy can be shown to occur only in hydrogen, the only atom wherein the electron moves always in a Coulomb field with potential energy $-e^2/r$. All orbitals of s type are expressible as a function $f_{ns}(r)$, where r is any radial co-ordinate. Such orbitals are therefore spherically symmetrical and for all values of n must be single. On the other hand, p functions are triple, and expressible in the forms

$$xf_{np}(r), yf_{np}(r) \quad \text{and} \quad zf_{np}(r).$$

Their symmetry in respect to the co-ordinate axes can only suggest that whatever form the potential field may take their threefold degeneracy will not be removed, although their associated energy will differ from that of the ns function. Similar arguments lead us to expect that the fivefold degeneracy of nd orbitals and the sevenfold degree of nf orbitals persist in the normal states of all elements.

In the years since 1913 the supreme task of explaining the electronic constitution of elementary atoms had progressed to a conclusion, and an attack upon the further problem of the chemical interaction of atoms became at last possible on a less speculative foundation than before: 'the present confusion' hopefully regarded as a passing phase by an eminent chemist in 1923 has been notably alleviated, if not entirely dispelled. To future generations Niels Bohr and Louis de Broglie will doubtless continue to appear as the great innovators, but it has been well observed that, given the actual development of their ideas which took place, the periodic system and no less the mechanism of chemical union would still remain inexplicable without Pauli's exclusion principle.

According to well-established thermochemical principles a

molecule is more stable than its isolated constituent atoms because large energies are always released in its formation from them

$$2H = H_2, \qquad \Delta H = 104 \text{ kcal.,}$$
$$2N = N_2, \qquad \Delta H = 225 \text{ kcal.,}$$
$$2H + O = H_2O, \qquad \Delta H = 220 \text{ kcal.}$$

In these examples ΔH is the energy released (in heat units) per gram-molecule formed. A convincing theory of chemical combination should at least indicate the source of such energy, even if it cannot evaluate it precisely. The vibration of the atoms about mean positions in a molecule implies attractive forces resisting elongation and repulsive forces resisting contraction, of the internuclear distance. The essential physical property of a stable molecule in thus possessing characteristic mean internuclear distances ought also to emerge from a theory of its formation. Further, the powerfully maintained stereochemical relations between the links formed by an atom ought to receive elucidation. Great advances have indeed been made in all these matters, but only the early stages come within our period to 1930.

In 1923 Sidgwick offered the significant suggestion that the chemical link, due according to the principle of Lewis to the sharing of a pair of electrons, might be implemented by the occupation of *binuclear* or two-centre orbits, the two-centred condition distinguishing a bonding or *molecular* orbit from a one-centre or atomic orbit.* After the successful application of wave mechanics to the one-centre problem of the hydrogen atom, it might have appeared possible to attack the two-centre case of H_2 by similar means, and a wave-equation formally representing the system is easily set up:

$$\nabla_1^2 \psi + \nabla_2^2 \psi + (8\pi^2 m/h^2) \, (E - V_1 - V_2 - V_{12}) \psi = 0.$$

ψ is here a function of the six co-ordinates of the two electrons: $\psi^2 dx_1 \, dy_1 \, dz_1$ gives the probability of finding electron 1 within a volume element round the point x_1, y_1, z_1, and $\psi^2 dx_2 \, dy_2 \, dz_2$ the probability of simultaneously finding electron 2 in a volume element round the point x_2, y_2, z_2. In the four terms expressing

* P. 140.

the potential energy $V_{1,2} = +e^2/r_{1,2}$ is the repulsion energy of the electrons. The unsurmountable difficulty that $r_{1,2}$ cannot be known until ψ is known renders the wave-equation inherently insoluble. All theoretical treatments of molecules by wave-mechanics have therefore been based on molecular wave-functions constructed algebraically from the known atomic functions of the combining atoms.

W. Heitler and F. London, in the first successful attempt to apply such a method to the hydrogen molecule, represented with great ingenuity the assumed equal sharing of the two electrons in the diatomic system.* Let $\psi_a(1)$, $\psi_b(1)$; $\psi_a(2)$, $\psi_b(2)$, be respectively the normalized (1s) wave-functions for electron 1 in atom a or in atom b, and the similar functions for electron 2. The probability of a containing 1 is measured by $\psi_a^2(1)$, and that of b containing 2 by $\psi_b^2(2)$. Hence the probability of these assignments occurring simultaneously is $\psi_a^2(1)\,\psi_b^2(2)$ and the corresponding function is $\pm\psi_a(1)\,\psi_b(2)$. For the alternative and equally probable assignment of 1 to b, and 2 to a we reach similarly the function $\pm\psi_a(2)\,\psi_b(1)$.

To form the representation of equal sharing when the atoms are contiguous these two functions were then combined to give a *molecular* function Ψ in which the constituent functions had each equal weight. For such combination there are two independent expressions, depending on the signs chosen for the product functions, viz.

$$\Psi_+ = [\psi_a(1)\,\psi_b(2) + \psi_a(2)\,\psi_b(1)](1/\sqrt{2})$$
$$\Psi_- = [\psi_a(1)\,\psi_b(2) - \psi_a(2)\,\psi_b(1)](1/\sqrt{2})$$

The factor $1/\sqrt{2}$ is required for the normalization of Ψ. Although ψ_a and ψ_b have identical algebraic forms their co-ordinates in the diatomic system are not the same. If the atoms approach each other along the z axis of rectangular co-ordinates conveniently centred on the mid-point of the internuclear distance r, then for a and b respectively the co-ordinates will be $x, y, z \pm \frac{1}{2}r$.

* 'Wechselwirkung neutraler Atome und homöopolare Bindung nach der Quantenmechanik. *Z. Phys.* **44**, 455 (1927).

In the diatomic system the total electronic potential energy is

$$V = -e^2\left[\frac{1}{r_{1a}} + \frac{1}{r_{2b}} + \frac{1}{r_{1b}} + \frac{1}{r_{2a}} - \frac{1}{r_{1,2}}\right],$$

where $e^2/r_{1,2}$ is the mutual repulsion energy of the electrons, but if electron 1 is originally in a and 2 in b, then, as we wish to calculate the *change* of energy as the atoms approach each other, the first two terms may be omitted from the operator H in the wave-equations for Ψ_+ and Ψ_-.

From the equation $H\psi = E\psi$ we have $\psi H\psi = E\psi^2$, and if ψ is normalized and the integration carried over all co-ordinate space, $\qquad \int \psi H \psi dv = E,$

since $\int \psi^2 dv = 1$ (p. 149).

Writing E_+ and E_- for the electronic energies associated with Ψ_+ and Ψ_- respectively, and noting that H is a linear operator,

$$E_+, E_- = \int (uHu + wHw)\,dv \pm \int (uHw + wHu)\,dv,$$
$$= E_{uu} \qquad\qquad \pm E_{uw}$$
$$(u = \psi_a(1)\,\psi_b(2), \quad w = \psi_a(2)\,\psi_b(1)).$$

On inserting the known 1s functions of the hydrogen atom in terms of the type of co-ordinates previously mentioned, E_{uu} and E_{uw} are calculated, necessarily as functions of the internuclear separation r. E_{uu} and E_{uw} then appear as both negative in sign, and $|\,E_{uw}\,| \gg |\,E_{uu}\,|$ for significant values of r:

$$|\,E_{uw}\,| \simeq 10\,|\,E_{uu}\,|;$$

moreover, $|\,E_{uu}\,|$ and $|\,E_{uw}\,|$ as well as $d\,|\,E_{uu}\,|/dr$ and $d\,|\,E_{uw}\,|/dr$ increase monotonously with decreasing internuclear separation. The effect of the approach of the two atoms on the electronic energies may be graphically presented thus:

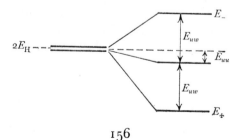

the 'splitting' of the originally coincident energy levels increasing with the closeness of approach of the atoms.

The Schrödinger equation thus yields the result that two hydrogen atoms held in contiguity may be subject to strong *attractive* forces (dE_+/dr positive, wave function Ψ'_+), or almost equally strong repulsive forces (dE_-/dr negative, wave-function Ψ'_-).

Since its first enunciation (p. 145) Pauli's exclusion principle had been given a more general form, viz. 'On the interchange of any two electrons within a system of bound electrons, the wave function $\dfrac{\text{remains unchanged}}{\text{changes sign}}$ when the interchanged electrons have $\dfrac{\text{opposed}}{\text{the same}}$ spins.

Noting that Ψ'_- changes sign with interchange of electrons 1 and 2, but Ψ'_+ does not, it is seen that two hydrogen atoms attract or repel each other depending on whether the electrons are antiparallel ($\uparrow\downarrow$) or parallel ($\downarrow\downarrow$) in spin. G. N. Lewis was gifted with a remarkable prescience when in 1923, before the discovery of electron spin, he emphasized the importance of the fact that molecules tend to be diamagnetic (p. 140).

If the electron spins are opposed the electronic energy becomes steadily more negative, and attractive forces stronger, as the internuclear distance is contracted. The electron energy, however, includes only the electron-proton and electron-electron interactions, but not the proton-proton repulsion, which steadily increases as the atoms converge, until at a definite value of r the attractive and repulsive forces balance: further contraction is opposed by a net repulsion, and retrogression to larger r by attractive forces. A fixed mean internuclear distance in the molecule is thus predicted by the theory.

Heitler and London extended their theory of the hydrogen molecule to explain why two atoms of helium, each with two electrons in the ground state $1s$, could only repel each other. By Pauli's principle the electrons in each atom must maintain opposed spins ($\uparrow\downarrow$, $\uparrow\downarrow$). Hence in the system He + He only exchanges with *parallel* spins are possible and only antisymmetrical wave-functions of the type Ψ_- in the case of hydrogen, can exist.

In their first joint paper and in subsequent independent communications Heitler and London laid stress on the differing nature of the two integrals which jointly determine the interaction energy between two hydrogen atoms. The first integrand, of type uHu, leading to the same interaction energy $-C = E_{uu}$ in the case of both Ψ_+ and Ψ_-, represents the calculation by wave-mechanical principles of the classical Coulombic interaction of electrons with protons and with each other. On classical principles the small negative energy of interaction $-C$, quite insufficient to stabilize a molecule, is all that could be predicted. The second integrand, which yields by far the major part of the interaction energy, $\pm J = E_{uw}$, is of the type uHw and is peculiar to quantum mechanical principles in that it depends on the *exchange of electrons*

$$u = \psi_a(1)\,\psi_b(2), \; w = \psi_a(2)\,\psi_b(1).$$

Heitler and London proposed the term *exchange energy (austausch energie)* for $\pm J$ and regarded its existence as the fundamental basis of the chemical interaction of atoms.

In Heitler and London's treatment of atomic interaction, exemplified in the hydrogen molecule and the inertness of helium, the crucial part played by electron spin was self-evident, and the assimilation of this concept into the general wave-mechanical method became an essential condition of the rapid development that followed upon their exposition.

During 1926, the year preceding the publication of Heitler and London's paper, W. Heisenberg* and independently P. A. M. Dirac† had demonstrated by very general mathematical methods that the wave-equation for any system of particles could be satisfied by only two types of wave-function— the *symmetrical* and the *anti-symmetrical*: a permutation of any two particles leaves the former unchanged but causes the second to change sign. No function not possessing one or the other of these characters is acceptable. Moreover, a system once

---

* *Z. Phys.* **38**, 411, June 1926.   † *Proc. roy. Soc.* A, **112**, 66, Aug. 1926.

'symmetrical' cannot by any spontaneous change become 'anti-symmetrical', or vice versa.

Now the exclusion principle in its first specialized form (1925, p. 145) proved that any system in which the particles are electrons and nuclei must be represented by an anti-symmetrical function. For such assemblies therefore symmetrical functions are excluded. It was appropriately left to Pauli* to show a mode by which this general result could be simply incorporated into wave-mechanical theory, and to give to the exclusion principle its final most general form—all (complete) wave functions of systems of electrons and nuclei are antisymmetrical to all permutations of two particles.

Without committing ourselves to the physical reality of either orbital or spin motions of a bound electron, we may still confidently assert that the concept of quantized orbital and spin angular momentum has meaning, for it leads to correct predictions of behaviour. Since a specific set of wave-functions (of spatial co-ordinates) is associated with a given orbital angular momentum chosen from the quantized values $(0, 1, \ldots, n) \, h/2\pi$ it is logical to assume that wave-functions (of a spin co-ordinate σ) are characteristic of electron spin. The angular spin momentum being restricted to the two values $\pm \frac{1}{2}(h/2\pi)$ there can exist only two spin functions α and β (of a co-ordinate σ) respectively corresponding to these values. A spin wave-function for two electrons 1 and 2 is assumed to submit to Heisenberg and Dirac's conditions: the only possible antisymmetrical form is

$$\Sigma_- = \alpha(1) \, \beta(2) - \alpha(2) \, \beta(1),$$

indicating opposed spins.

Now we have seen that only the symmetrical *orbital* function Ψ'_+ leads to binding energy between two hydrogen atoms. Nevertheless, the generalized exclusion principle requires that the *total* wave-function (orbital with spin) shall be antisymmetrical. Such a total function is clearly

$$\Psi'_+ \, \Sigma_-,$$

* 'Zur Quanten mechanik des magnetischen Elektrons', *Z. Phys.* **43**, 601, Mar. 1927.

since (symmetrical) × (antisymmetrical) ≡ (antisymmetrical). Lacking the appropriate spin function we could only choose the antisymmetrical function Ψ_- which gives not bonding but repulsion, between the atoms. In an independent paper 'Zur Quantentheorie der homöopolaren Valenzzahlen'* London endeavoured to develop a general theory of valency, using the theory of the hydrogen molecule as a prototype for other simple combinations, just as the hydrogen atom had provided a starting-point for the elucidation of the constitution of other atoms. In opening the paper London claims that the very existence of valency appears to depend upon 'austauschenergie' and the concept of electron spin:

Gäbe es in der Wechselwirkungsenergie nur die Coulombsche Wechselwirkung der räumliche Ladungswolken der Einzelatome, so gäbe es keine Valenzzahlen, alle Materie stürzte chaotische ineinander in den Zufällig vorliegenden Masseverhaltnisse. Die Diskontinuitäten der Chemie, welche in Gestalt des Gesetzes der multiplen Proportionen seinerzeit die Ideen der Atomistik lebendig gemacht hatten, beruhen wie man hier sieht, nicht so sehr auf dem Atomismus der Materie.† Man erhielte eine regellose Anziehung, wenn nicht zu dem Coulombsche Bestandteil der Wechselwirkungsenergie als ein typisch quantenmechanischer Effekt die sogenannten Austauschenergien treten würden, welche nun auf eine sehr characteristische Weise die Verhaltungsweisen zweier Atome regeln.

Gäbe es keinen Elektronendrall, so würde das Pauli-Prinzip nur die antisymmetrische Lösung mit Abstossung zulassen, und es gäbe keine homöopolare Bindung. Die Tatsache der homöopolare Chemie scheint im Zusammanhang mit dem Pauli-Prinzip ausschliesslich auf dem Vorhandensein des Dralls zu beruhen.

(*Were there only the Coulomb interaction (calculated albeit according to wave mechanical principles) between atomic charges, no definite numerical valency would be possible: matter would agglomerate chaotically in any fortuitous proportions. The discontinuities typical of chemistry, which in the guise of the law of multiple proportions formally inspired the revival of an atomic theory, are now seen not to rest entirely upon such a concept. Only an unregulated attraction results unless on the Coulombic part of the interaction the quantum-mechanical effect of 'exchange energy' is superposed to control and regulate characteristically the atomic interaction.*

If there were no electron spin the Pauli principle would permit only the antisymmetrical wave-mechanical solution (of atomic interaction), with repulsion between atoms: covalent binding could not arise. The fact of (covalent) chemical union appears in relation to the Pauli principle to rest exclusively on the existence of electron spin.)

* *Z. Phys.* **46**, 455 (1927).   † Cf. Berzelius, p. 26.

London also sees the exclusion principle as the cause of the property of saturation which confers upon valency its unique character:

Eine freie Valenz, welche durch eine entsprechende freie Valenz eines anderes Atoms abgesättigt ist, scheidet wirkliche für alle anderen Prozesse aus; sie kann nicht (wegen des Pauli-Verbotes) etwa noch eine weitere Valenz absättigen.

(*A free valence, saturated by a corresponding free valence in another atom, is excluded from all further (interaction) processes: because of the exclusion principle it cannot saturate any other valence.*)

The valency electrons of any atom, i.e. those capable of taking part in the formation of interatomic links, must lie under no restraint in regard to their spins. Therefore, under the 'aufbauprinzip', such active electrons can only be found among those least strongly bound on the periphery of an atom, where alone it is possible for orbitals to contain single electrons with uncoupled spins. The valency of any atom is just the number of such 'lone' electrons it contains. As examples we may take the elements preceding neon: in an atom of nitrogen the outermost shell containing $Z' = 5$ electrons could be assigned the configurations (*a*) or (*b*), of which the latter is the more stable (ground) state:

Nitrogen $Z' = 5$	2s	2p			Valency
	0	-1	0	1	
(*a*)	2	2	1	.	I
(*b*)	2	1	1	1	III
(*c*)	1	1	1	1	IV

When one electron is removed by ionization the quadrivalent configuration (*c*) results, and leads to the formation of the ammonium ion $(NH_4)^+$. In oxygen ($Z' = 6$) there can be only two unrestricted electron spins, in fluorine ($Z' = 7$) one, and in neon ($Z' = 8$) none:

	2s	2p			Valency
	0	-1	0	1	
Oxygen, $Z' = 6$	2	2	1	1	II
Fluorine, $Z' = 7$	2	2	2	1	I
Neon, $Z' = 8$	2	2	2	2	O

For the heavier members of the Groups V, VI and VII a range of valencies becomes possible, owing to the greater number of orbital types in their 'valency shells':

	ns	np			nd					Valency
	o	-1	o	1	-2	-1	o	1	2	
Group V	2	1	1	1	.	.	.	.	.	III
(P, As, Sb)	1	1	1	1	1	.	.	.	.	V
Group VI	2	2	1	1	.	.	.	.	.	II
(S, Se, Te)	2	1	1	1	1	.	.	.	.	IV
	1	1	1	1	1	1	.	.	.	VI
Group VII	2	2	2	1	.	.	.	.	.	I
(Cl, Br, I)	2	2	1	1	1	.	.	.	.	III
	2	1	1	1	1	1	.	.	.	V
	1	1	1	1	1	1	1	.	.	VII

In thus presenting a physical basis for valencies previously established in purely chemical operations, London introduced the important proviso that in assessing valency from electronic configuration *excited* states neighbouring the ground-state must be included when they can give rise to an increase of valency. He regarded this as particularly necessary in the case of carbon, which in its ground state can be only bivalent:

Carbon	$2s$	$2p$			
$Z' = 4$	o	-1	o	1	
	2	1	1	.	Ground-state
	1	1	1	1	4-valent state

London was in this way able to provide an explanation both for the 'polyatomicity', changing by steps of two, discovered by Frankland more than sixty years previously, and for the constant maximum valencies of carbon (IV, as in N^+), nitrogen (III) and oxygen (II) so confidently maintained by Kekulé. By adopting the suggestion of Hantzsch* that the constitution of anhydrous nitric acid was $(HO)^- (N^+O_2)^+$ he hoped to discredit the still

* *Z. Elektrochem.* **29**, 221 (1923).

prevailing opinion among chemists that nitrogen could exhibit quinquevalency. The resultant distribution of electric charge in the inert elements, such as neon, was known to be spherically symmetrical about the nuclei. London put forward the view that since the four valency electrons in C and N^+ occupied singly all the orbitals doubly occupied in neon, these two species must also possess spherical symmetry, a condition which accounted for the maximum (tetrahedral) symmetry of CH_4 and $(NH_4)^+$ with their substitutional derivatives.

Towards the end of his paper London seeks to throw light on the conditions which determine whether two atoms A and B become linked by the purely electrostatic interaction of ions, e.g. A^+B^-, or by the 'homöopolare' binding (covalent in Langmuir's terminology) due to shared electrons. In almost every known case the formation of (isolated) ions A^{z+} and B^{z-} from the neutral atoms A and B is strongly endothermic: $E_{ions} = I_A + E_B$, $(I_A + E_B) > 0$, where I_A is the ionization potential of A, and E_B the electron affinity of B. If the isolated ions are allowed to converge and attract each other, supposedly under an inverse square law, then at an interionic distance r_s given by

$$I + E = (z_+ z_- e^2)/r_s,$$

the ionic system regains the energy level of the original neutral atoms. London suggests that when r_s is much greater than the actual interatomic distance in the molecule or crystal of AB then an ionic linking is indicated.

	r_s (A)	A – B in crystal (A)
KF	60	2·66
KCl	29	3·14
KBr	17	3·28
KI	11·8	3·51

		A – B in molecule
HF	1·51	0·94
HCl	1·47	1·28
HBr	1·42	1·42
HI	1·37	(1·50)

11-2

When r_s is comparable with (or even less than) the interatomic distance a more complicated situation arises through the quantum-mechanical interaction between the (degenerate) ionic and covalent systems. London writes, somewhat enigmatically, of 'einer mehr oder weniger homöopolaren Verbindung', but so foreshadows a later explanation of the electric polarity of heteronuclear molecules.

As attention became increasingly drawn towards theories of molecule formation it was found necessary to devise a means of describing the electronic state of a molecule. The method long used for atoms by spectroscopists was as follows:

Angular momentum
(arbitrary axis)

	Individual electrons			Atomic resultant		
l	0	1	2	0	1	2
Symbol for electronic state	s	p	d	S	P	D	Symbol for atomic state

This scheme was imitated for diatomic molecules, which, unlike atoms, possess a natural axis of quantization in the internuclear line:

Diatomic molecules. Angular momentum
(about the internuclear line)

	Individual electrons					Molecular resultant		
l	0	1		2		0	1	2
Symbol for electronic state	σ	π_+	π_-	δ_+	δ_-	Σ	Π	Δ	Symbol for molec state

When for an atom or molecule the resultant spin is $\Sigma s = S$, the number $2S + 1$ (the multiplicity) is inserted as a left-hand superscript, e.g.

$$^1S, \, ^3P; \, ^3\Sigma, \, ^2\Pi.$$

The states π_+ and π_- (as well as δ_+ and δ_-) differ only in the direction of motion about the internuclear axis: such molecular states are therefore equivalent and degenerate, like p and d atomic states.

In a short paper in 1928* Heitler paid the following tribute to the concepts of G. N. Lewis

Die Theorie der homöopolaren chemischen Bindung zweiatomige Moleküle
...ergab eine völlig Äquivalenz der Theorie mit der Elektronenpaarvorstellung von Lewis: zwei freie Leuchelektron, die zu zwei verschiedenen Atomen gehören, können vermöge ihrer Austauschenergien eine Auziehung der Atome bewirken.

(*The (quantum-mechanical) theory of covalent binding in diatomic molecules is completely equivalent to Lewis's electron-pair concept: two free 'valency' electrons, belonging to two different atoms, can, in virtue of their 'exchange energy', cause an attraction between the atoms.*)

In spite of the striking and welcome authentication of Lewis' qualitative principles, the number of important examples of their transgression was sufficient to encourage the more sceptical to withhold full confidence in the current theory. Lewis (1916), and by implication also London, laid great stress on the prevalence of molecules containing an even number of electrons, and Lewis (1923) had seen the counterpart of nearly universal electron sharing in the prevalence of diamagnetism among compounds: but in presenting a model of the 'even' molecule of oxygen in 1916 he had ignored its strong paramagnetism, suspected by Faraday in 1847, and proved by Dewar (employing the liquefied gas) in 1892.† The 'odd' molecules dismissed by Lewis comprise some very stable and simple molecules, such as nitric oxide, NO, and nitrogen dioxide, NO_2. Indeed these are far more stable than the 'even' molecules N_2O_3 and N_2O_4 to which they give rise by association at lower temperatures. In 1927 Ellsworth and Hopfield‡ studying the various molecular spectra to which oxygen gives rise, proved that the 'odd' molecular ion O_2^+, formed from O_2 by ionization, is considerably more stable, i.e. has a higher energy of dissociation, than its parent.

The chemist might have felt much disappointment that in the long-standing valency problems involved in molecules such as NO, NO_2, and CO he received so little help from the progress of

* *Z. Phys.* **51**, 805 (1928). † *Proc. roy. Soc.* **50**, 247 (1892).
‡ *Phys. Rev.* **29**, 79 (see p. 171)

11-3

modern theories. It is true that Langmuir had written in 1919*
'the octet theory indicates that the number and arrangement of
electrons in N_2, CO and the cyanide ion CN^- are the same', but
beyond noting the exceptional similarity in the physical properties of nitrogen and carbon monoxide he had pursued the
question no further. In 1927 both Sidgwick† and London‡ (see
p. 162) were disposed to regard carbon as bivalent in its monoxide, but in view of the chemical promiscuity of carbon it
seems incomprehensible that a molecule $C{=}O$ should be so
extremely stable while CH_2 and CCl_2 are undiscoverable.

During the years 1926–1928, that is contemporaneously with
Heitler's and London's publications of what came to be known
as the 'valence bond' method, another way of approach to
molecular problems, to be later called in distinction the method
of 'molecular orbitals' had been developed, mainly by the
inspiration of the spectroscopist F. Hund.§ The method
achieved the outstanding distinction of affording one comprehensive explanation not only of the types of molecule already
treated by the 'valence bond' method but also of the other more
unusual examples mentioned above. Hund's basic idea involved a certain disregard of the chemical tradition of the con-
tinuing existence of atoms in a molecule, and its replacement by
a real 'aufbau' process essentially similar to that by which
atomic constitution had been so successfully established. The
two nuclei of a diatomic molecule AB, ultimately with total
charge $Z_A + Z_B$ are supposed to be held in contiguity while the
$Z_A + Z_B$ electrons are introduced into the available molecular
states in order of energy and under the operation of the exclusion principle.

Im folgenden‖ soll auch versucht werden, über das Zweizentrensystem mit
beliebigem Zentrenabstand Näheres auszusagen, insbesondere, ob und

* *J. Amer. chem. Soc.* **41**, 1543.
† *Electronic Theory of Valency* (Oxford, 1927), p. 272.
‡ *Z. Phys.* **46**, 455 (1927).
§ 'Zur Deutung der Molekülspektren', *Z. Phys.* **40**, 742 (Nov. 1926); **42**, 93
(Feb. 1927); **43**, 804 (May 1927); **51**, 759 (Oct. 1928); the last paper offers a con-
spectus of the developed theory. R. S. Mulliken, *Phys. Rev.* **32**, 186 (1928) gives a
valuable exposition and commentary on Hund's papers.
‖ *Z. Phys.* **51**, 759 (1928).

wieweit sich ein *Aufbauprinzip* aufstellen lässt, das dem für Atome aufgestellten analog ist. Wir denken uns dabei die beiden Kerne in beliebigem Abstand festgehalten, fügen diesem System schrittweise je ein Elektron zu unter geeigneter Erhöhung der Ladungen der Kerne, und untersuchen den jeweils dadurch hergestellten Quantenzustand.

The central necessity is some method of pre-determining at least the order of the binding energy in the molecular states. There was then (and is now) little hope of a general calculation of such energy levels by quantum-mechanical theory, but Hund observed that the molecular states must be interposed between two other sets of states, both of which are exactly known from spectroscopic data: namely those of the constituent atoms, and those of a (real) atom of atomic number $Z_A + Z_B$, e.g. $N + N$ and Si; $O + O$ and S; $N + O$ and P; $C + O$ and Si. States are imagined to pass with a mathematical continuity from the un-combined atoms, through the molecular set, to the 'united atom'.

States of the 'united atom' of the same quantum numbers (e.g. $2s$, $2p$, $3s$, etc.) as are also to be found in the constituent atoms must lie deeper on a common energy scale because of the increase of the nuclear charge. Owing, however, to the fact that such levels are duplicated in the two atoms and single in the 'united atom' one half of the atomic levels must in the 'united atom' be 'promoted' to levels of higher quantum numbers, a change which will in general involve a *rise* of energy. We can therefore expect that the predictable molecular states will consist of two equal sets, the one comprising energy levels deeper than those in the two atoms, and the other formed of levels higher than those in the atoms. The quantum-mechanical 'splitting' of atomic levels as two atoms converge, given prominence in Heitler's and London's treatment, becomes again recognized as a general phenomenon in Hund's method, but in that method it is an essential requirement that when more than two electrons are at disposal electrons will of necessity be found in the molecule in both sets of levels, exerting a *repulsive* effect in the promoted levels, and a binding effect in the others.

Two or more atoms can form a molecule, if the energy of the complete molecule is less than the sum of the energies of the separate atoms. Whether that is the case can be decided by a very complicated and laborious calculation, using the new quantum mechanics. We cannot satisfactorily use such a calculation; in order to understand chemical binding, we look for simple properties of the atoms which lead us to expect a positive or a negative result from the calculation. It may be that there are different conditions, enabling atoms to form compounds, which can work against each other. It is in that case not possible, in a single case, to predict whether there is binding or not; nevertheless we can say that we understand binding.†

The detailed correlation of levels is effected largely by considering the symmetry properties of the relevant wave-functions: and in many cases the assignment of levels to a molecule can be checked by experimental data from molecular spectra. For pairs of atoms chosen from Series II of the Periodic System (lithium-neon) the correlation scheme between uncombined atoms and diatomic molecules takes the following general form:

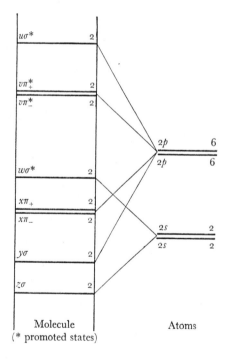

† Hund, 'Chemical Binding', *Trans. Faraday Soc.* **25**, 646 (1929).

Molecular states of the σ type are single and can each contain a maximum of 2 spin-coupled electrons: π states are double, comprising the two degenerate states π_+ and π_- (p. 164), which together contain a maximum of 2 pairs of coupled electrons. For all atoms from lithium onwards it is assumed that the paired $1s$ (K) electrons lie so deep in the atom that their interaction in a molecule is negligible: in Lewis' terms they with the nuclei constitute the atomic 'kernels' for Series II atoms.

	$\mathcal{Z}-4$	$z\sigma$	$y\sigma$	$x\pi$	$w\sigma^*$	$v\pi_+^*$ $v\pi_-^*$	$u\sigma^*$	State	Bond order
N_2, CO	10	2	2	4	2	.	.	$^1\Sigma$	3
NO, O_2^+	11	2	2	4	2	1	.	$^2\Pi$	$2\frac{1}{2}$
O_2	12	2	2	4	2	1+1	.	$^3\Sigma$	2
F_2	14	2	2	4	2	4	.	$^1\Sigma$	1
Ne+Ne	16	2	2	4	2	4	2	—	0

The disposable electrons in N_2, CO, F_2 (and 2Ne) fall uniquely into closed groups, as do ten of the twelve electrons in O_2: for the two remaining electrons the degenerate states π_+ and π_- have to be considered. All chemically active atoms must necessarily contain in their peripheral shells an insufficient number of electrons to constitute a complete set of closed groups: as for example in the nitrogen and oxygen atoms there are only three and four electrons respectively to be accommodated in the three (degenerate) $2p$ orbitals. From atomic spectra it is observed without exception that the most stable arrangement and therefore that of the ground-state is the one of maximum multiplicity. This rule, essentially involved in London's theory of valency,[*] was first formulated by Hund[†] and has been set upon a firm quantum-mechanical basis, justifying its extension to 'molecular' electrons. Hence, of the three possible assignments in O_2

$$\pi_{+\,-}(\uparrow\downarrow),\; ^1\Delta;\quad \pi_+(\uparrow)\,\pi_-(\downarrow),\; ^1\Sigma;\quad \pi_+(\uparrow)\,\pi_-(\uparrow),\; ^3\Sigma;$$

[*] P. 161
[†] 'Linienspektren', 1927, p. 124.

the last is the lowest and most stable. Hund's scheme therefore necessarily predicts the magnetism of molecular oxygen, and ascribes to it the magnitude arising from two parallel electron spins. The molecules of nitrogen and fluorine are diamagnetic and the molecule of oxygen paramagnetic as a necessary consequence of their being in their most stable states.

In nitric oxide the alternative assignments $\pi_+(\uparrow)$ or $\pi_+(\downarrow)$ correspond to the doublet indicated by the symbol $^2\Pi$. In one state the spin and orbital momenta lie in the same direction (resultant $\frac{3}{2}h/2\pi$) and in the other oppose each other (resultant $\frac{1}{2}h/2\pi$). It is a remarkable property of electron spin that a (hypothetical) non-spinning electron, moving in a 'p' orbit in the same direction co-axially with the spin, produces the same magnetic field as that due to the spin alone, although the angular momentum of the latter is only half that of the (non-spinning) electron. Consequently the (lower) state $^2\Pi_{\frac{1}{2}}$ is diamagnetic while the state $^2\Pi_{\frac{3}{2}}$ is paramagnetic. Owing to the very feeble coupling of the spin and orbital momenta these two states differ very little in energy (roughly $kT/2$ per molecule), so that at ordinary temperature nitric oxide exists in a state of equilibrium between the states, which at low temperatures shifts strongly in favour of the diamagnetic state.

Although Hund's approach to molecular constitution does not lend itself to transcription into traditional graphic formulae it is striking that the expression $(n - n^*)/2$, in which n is the total number of binding electrons, and n^* the total number exerting repulsive action ('anti-binding' electrons), gives correctly the 'bond order' or the number of bonds traditionally assumed by chemists (and in the valency–bond theory of London) in the molecules N_2, O_2, F_2 (and $Ne+Ne$).[†] Applied to carbon monoxide this interpretation leads to the graphic formulae $C{\equiv}O$. As Langmuir suggested (1919, p. 166) the formulae of CO and N_2 are alike; each contains a triple link between the atoms. The molecular theories of Heitler and of London had appeared to require that in the formation of diatomic molecules

---

[†] J. E. Lennard-Jones appears to have first noticed this relationship, cf. *Trans. Faraday Soc.* **25**, 684 (1929).

each constituent atom should supply an equal number of binding electrons, which revert equally to the atoms upon dissociation of the molecule. No such limitation exists in Hund's molecular 'aufbau' process, which in the example of CO requires that of the ten electrons involved four pass to carbon and six to oxygen on dissociation. In terms of the formula $C\equiv O$ this could mean that of the three common pairs of electrons one pair is supplied entirely by the oxygen atom. To chemists such an unequal sharing of electron pairs was nothing novel, for Sidgwick in 1923 (p. 137) had already given a comprehensive and convincing explanation of 'coordination compounds' on a basis of the 'co-ordinate link' between a 'donor' and an 'acceptor' atom, and in his symbols carbon monoxide appears as $:C \leftrightharpoons O:$.

A fractional bond order such as appears for nitric oxide and the molecular ion O_2^+ is not easily translated into traditional terms but that it has physical significance is supported by the relation with the observed energies of dissociation

	Energy of dissociation (kcal.)	Bond order
$N_2 \to 2N$	225	3
$NO \to N+O$	150⎫	$2\frac{1}{2}$
$O_2^+ \to O+O^+$	150⎭	
$O_2 \to 2O$	117	2

REFERENCE

The Electronic Theory of Valency. N. V. Sidgwick. (Clarendon Press, Oxford, 1927.) (A survey, mainly in the field of physical and inorganic chemistry, up to the eve of the application of wave-mechanical principles.)

CONCLUSION

The decay of the barriers so long separating chemical and physical science and the focusing of physics in the guise of the new quantum mechanics upon the great empirical structure of chemistry, led, not earlier than 1927, to an understanding of the periodic system of classification and of why two hydrogen atoms may, given certain conditions, converge upon each other with the release of a large energy and the formation of a stable molecule. Chemists had so long studied, and with great success, the vast range of facts coming into their cognizance that the absence of a physical basis had come to be regarded as almost irrelevant, or at least outside the proper scope of chemical science.

In the years immediately succeeding these first contributions the new 'chemical physics' was content to try to catch up with the chemical doctrine of valency, and as a reflection upon this period we may recall the sober opinions of Heisenberg previously quoted (p. 125). By about 1930 so much progress had been made that in the words of an eminent physicist: 'It may now be said that the chemical theory of valency is no longer an independent theory in a category unrelated to general physical theory, but just a part—one of the most gloriously beautiful parts—of a single self-consistent whole, that of non-relativistic quantum mechanics.' Perhaps a chemist may take some pride in that Professor R. H. Fowler felt disposed to add 'I have at least sufficient chemical appreciation to say rather that quantum mechanics is glorified by this success than that now "there is some sense in valencies" which would be the attitude of some of my friends'.*

In recent years the new theories have given rise to concepts transcending the classical chemical principles, and especially the graphic formulae traditionally used to embody them. Perhaps

* R. H. Fowler, A Report upon Homopolar Valency and its Quantum-mechanical Interpretation (British Association, 1931).

the most revolutionary of these advances is that concerning the 'delocalization' of electrons involved in binding atoms, so that the same electrons may serve to link a considerable group of atoms. Since such spatial expansion of electron function is especially well exemplified in the hydrocarbon benzene, the more conservative branch of chemistry has not been exempt from the invasion of such new physical concepts. It appears that the time has now arrived when it may be not unjustly claimed 'the more precisely the interactions between atoms are known, the less need is there to introduce the idea of valency. If it were possible to work out in practice...the interactions and the energy of any system of atoms the need for introducing the concept of valency would not arise.'*

The latest views upon the constitution of benzene originate from Hund's concepts of molecular orbitals. A molecule of benzene, C_6H_6, must contain $(6 \times 4) + 6 = 30$ valency electrons: of these 24 are employed in 12 electron pairs localized between pairs of carbon atoms or between carbon and hydrogen atoms respectively. One valency electron then remains in an atomic orbital of each carbon atom, the six such singly occupied orbitals being all equivalent. In principle these electrons could be utilized in the benzene molecule in two ways: they might form binding pairs between alternate carbon atoms, and so generate the molecule represented by the Kekulé formula, with three double linkages. But if the carbon atoms all lie in one plane, which is known to be the case in benzene (but not in cyclo-octatetrene) a different allocation becomes possible, which leads to a much more stable condition of the molecule. The six equivalent carbon orbitals may combine into an equal number of orbitals involving *all the six carbon nuclei*: and, as in the case of diatomic molecules, falling into a range of associated energies, in such a way that three of the molecular orbitals are binding and three anti-binding (cf. p. 167). Since there are just six electrons to occupy the orbitals they may all pass (by Pauli's principle) in pairs into the binding states.

* W. G. Penney (now Sir Wm. Penney), *The Quantum Theory of Valency* (London, 1935), p. 42.

In the resulting structure, which is some 40 kcal. more stable than the first proposed by Kekulé, the bindings between carbon atoms are all equivalent and each stronger than a single bond. Kekulé's later suggestion of 'mobile' double bonds, Baeyer and Armstrong's centric formula, and especially the theory of Thiele can all be considered to lie in the direct line of descent of this modern culmination. But the future is still faced with a further large task, for the characteristic delocalized binding can exist only in a molecule of benzene isolated from reagents, the influence of which has yet been little explored. Nothing in any more recent developments has detracted from Baeyer's thesis, that benzene under the influence of reagents (or of the presence of substituents) exists in a state lying between that of 'ideale Benzol', now supposed to contain six completely delocalized electrons, and that of the Kekulé formula, with electrons completely localized into the structure of a cyclic tri-olefine: it is certainly in the latter 'limiting state' when reacting with ozone.

APPENDIX

Symbols and constants (Chapter 8)

c Velocity of light $(2 \cdot 9979 \times 10^{10}$ cm. sec.$^{-1})$

e Electronic charge $(4 \cdot 803 \times 10^{-10}$ e.s.u.$)$

h Planck's constant $(6 \cdot 625 \times 10^{-27}$ erg. sec.$)$

m Electronic mass $(9 \cdot 108 \times 10^{-28}$ g.$)$

ν (Absolute) frequency (t^{-1})

λ Wave length

$\nu/c = 1/\lambda$ Wave number (cm.$^{-1}$)

ψ Wave function (atomic)

Ψ Wave function (molecular)

Hydrogen spectra and Rydberg's constant R

General formula $\dfrac{1}{\lambda} = R\left[\dfrac{1}{n_1^2} - \dfrac{1}{n_2^2}\right]$

$n_1 = 1, 2, 3, 4, 5; \quad n_2 = (n_1 + 1), (n_1 + 2), \quad$ etc.

Different series of spectral lines are given by combining any one value of n_1 with the members of the set n_2. For the lines of the visible (Balmer) spectrum $n_1 = 2$, and $n_2 = 3, 4, 5$, etc.

INDEX

INDEX